T0177235

THE LITTLE BOOK OF
WEATHER

THE LITTLE BOOK OF
WEATHER

With color illustrations by Claudia Myatt

ADAM SCAIFE

PRINCETON UNIVERSITY PRESS
PRINCETON AND OXFORD

Published in 2024 by Princeton University Press
41 William Street, Princeton, New Jersey 08540
99 Banbury Road, Oxford OX2 6JX
press.princeton.edu

Library of Congress Control Number 2024932726
ISBN 978-0-691-25999-4
Ebook ISBN 978-0-691-26016-7

Typeset in Calluna and Futura PT

Printed and bound in China
1 3 5 7 9 10 8 6 4 2

British Library Cataloging-in-Publication Data is available

This book was conceived, designed, and produced by UniPress Books Limited

Publisher: Jason Hook
Managing editor: Slav Todorov
Creative director: Alex Coco
Project development and management: Ruth Patrick
Design and art direction: Lindsey Johns
Copy editor: Caroline West
Proofreader: Robin Pridy
Color illustrations: Claudia Myatt
Line illustrations: Ian Durneen

IMAGE CREDITS:
Alamy Stock Photo: 51 Muhammad Mostafigur Rahman; 76 INTERFOTO/
History; 109 Jochen Tack/imageBROKER.com GmbH & Co. KG; 130 Album.
iStock: 22 FrankRamspott; 29 livetalent; 118 ugurhan; 124 Bob Gwaltney
Photography; 144 oversnap. **Nature Picture Library:** 68 David Noton.
Shutterstock: 10 Japan's Fireworks; 17 glenrichardphoto; 38 Hannah Stanbury;
45 John D Sirlin; 52 Mykola Mazuryk; 58 Alex Segre; 85 Paramonov Alexander;
88 Tobetv; 90 Andreanicolini. **Other:** 70 Chief Photographer's Mate (CPHoM)
Robert F. Sargent (National Archives and Records Administration); 97 © Stephen
Conlin 1986. All Rights Reserved. Research by Peter J. Byrne TCD; 98 U.S. Army;
137 NASA/GSFC/SDO; 151 NASA, ESA, STScl, A. Simon (Goddard Space
Flight Center), M.H. Wong (University of California, Berkeley), and the OPAL team.

CONTENTS

INTRODUCTION

From minor events like changing weekend plans to major life-changing weather disasters, the weather has a profound and irreversible effect on our lives, and it affects everything from the food we eat to the energy we need to keep warm. Although it might appear bewilderingly complex at first sight, weather is also amenable to scientific investigation and understanding. Scientific laws govern the weather and although it is far from complete, we now have a level of understanding that allows us to explain and predict important aspects of weather at lead times of days, months, or even years ahead. Our ability to simulate and predict the future using high-performance computers is an unsung triumph of science. This book tries to cover all these aspects and a good number of associated topics, ranging from weather-related turning points in history to the latest developments in forecasting that extend well beyond the usual daily forecast.

EARLY DAYS

As a child, I wasn't one of those kids with a weather station in the backyard, measuring temperature and rainfall, but I was always interested in the natural world and how things worked. From my early school days, I was drawn to natural history, science, and math. I was fascinated by insects with their alien forms and behavior, and fossils became another great obsession that I am still afflicted with today. You'll find examples in *The Little Book of Weather* that connect to all these topics. However, it was physics, with its rigorous descriptions of what we see around us and its apparently mysterious predictive power, that really captured my attention.

One of the few pieces of career advice I received was a passing comment from a physics professor giving a university lecture. While describing the Coriolis force on the blackboard he remarked over his shoulder, "They do this sort of thing at the Met Office; some of you might like to consider working there." I did not realize it at the time but this was a perfect forecast. After studying natural sciences at Cambridge University, I edged toward meteorology with a Master's degree in environmental science. A PhD and a position at the UK

Meteorological Office (now simply the Met Office) cemented my link with weather. *The Little Book of Weather* selects stunning examples of dramatic weather phenomena and weather effects from all over the world and from over 30 years of experience as a research scientist.

ABOUT THIS BOOK

Our knowledge of weather and its impact on the world is expanding rapidly and this is reflected in the choice of topics in *The Little Book of Weather*. There is a whole spectrum of ideas and subjects, ranging from insights into the mechanisms behind extreme weather phenomena and forecasts, to dramatic examples of where weather is thought to have changed the course of history or even altered the evolution of life on Earth. It is also now at the core of one of our most pressing concerns: climate change. As the Earth warms, weather events are changing and *The Little Book of Weather* takes a level-headed look at some of the key aspects of climate change and what we can expect in the future.

Whatever your passions, I hope there are sections in this book that will capture your enthusiasm and imagination and lead you to look further into our ever-changing weather.

Adam Scaife

THE ATMOSPHERE IS A FLUID

The next time you sit by a stream, take a closer look at the patterns and swirls in the water. You'll soon see waves traveling along the surface, with their regular ups and downs progressing along at a near constant rate across the water.

You'll also notice swirling eddies with a dimple in the middle where the water is being pushed outward by its circular motion. If it is turbulent enough, you'll see water upwelling and spreading outward in broad, shallow mounds on the surface. What you are looking at are analogs of weather systems in the atmosphere.

ROTATION AND WEATHER

To take all this further, we can even compare the atmosphere to the water in a bathtub: if we make our bathtub circular, rotate it slowly on a turntable and add an ice bucket in the middle to represent the cold polar regions, then all of a sudden it really does start to look like the atmosphere. Weather-like features spontaneously appear and start to swirl and propagate around just like the weather systems you see in a satellite image from space. Not only cyclones but even a jet stream forms in the water as the fluid obeys Sir Isaac Newton's laws of physics governing both it and the Earth's atmosphere.

ANALOGIES

Although at school we're taught to remember the distinction between gases and liquids, as far as nature is concerned, the atmosphere is just a shallow fluid trapped at the surface of the Earth by gravity, and it follows the same laws of physics as the water in a stream and the tea in your teacup. Of course, there are some differences: the air is very "thin," so that its internal friction or viscosity is small enough to ignore, and then there's the all-important rotation of the Earth and its near spherical shape (in fact, it's slightly hamburger-shaped). But the swirling eddies are like cyclones, the upwelling mounds of water are like thunderstorms, and those ripples on the surface are like the striped clouds we sometimes see stretching across the sky as the air inside rises and falls.

← These whirlpools and currents are in the ocean, but they are direct analogs of the circulating weather systems in our atmosphere, which also behaves as a fluid.

PRESSURE

Perhaps the most familiar of all weather elements is atmospheric pressure. Just as the pressure increases as you dive to the bottom of a swimming pool due to the weight of water above, so atmospheric pressure is simply due to the weight of the air above your head. This is why the pressure drops as you ascend a mountain and rise through the atmosphere—there is less air above your head and therefore less weight.

Atmospheric pressure is familiar to us all from television weather forecasts. Almost everywhere in the world, a drop in the reading on the barometer signals wet weather, or perhaps even an incoming storm, since all cyclones have low pressure at their center. Pressure measurements were the main source of information in early weather forecasts by Admiral FitzRoy (1805–1865), the captain of HMS *Beagle* (the ship on which Charles Darwin sailed around the world). He was acutely aware of the loss of lives and ships at sea due to bad weather, so he set up a series of barometers at points around the United Kingdom to make the first serious attempts at weather forecasting.

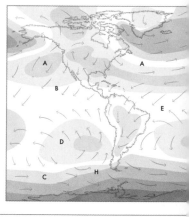

→ World pressure map showing the average sea level pressure across the globe (red = high pressure, blue = low pressure). Average 10-m winds show the midlatitude westerly jet streams in the North Pacific, North Atlantic, and Southern Hemisphere and the easterly trade winds in the tropics.

A westerly winds
B northeast trade wind
C prevailing westerly
D southeast trade winds
E doldrums
F northeast monsoon
G northwest monsoon
H roaring forties

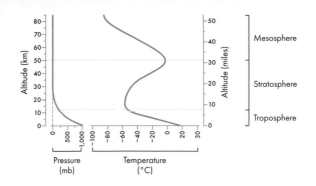

↑ Change in pressure and temperature with altitude. Pressure decreases exponentially with altitude whereas temperature decreases with altitude in the troposphere and the mesosphere, but increases with altitude in the stratosphere due to the warming effect of the ozone layer.

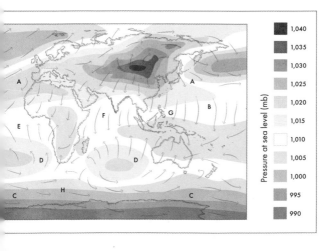

TEMPERATURE

What is temperature? What makes something feel hot or cold? We rarely stop to think about these questions but they are fundamental to the behavior of the weather. Temperature is simply a measure of the energy of motion of the molecules in the air. The faster a molecule is traveling through the air, spinning, vibrating, and jostling into its neighbors, the higher the temperature.

All this energy has to come from somewhere, and for the weather that energy source is the light from the Sun. Physicists discovered that energy is neither created nor destroyed, only converted from one form to another, so all that energy also has to go somewhere, and for the atmosphere it is ultimately sent back into space by the emission of heat and light.

TEMPERATURE SCALES

Several scales are commonly used for temperature and if you want to convert from Celsius to Fahrenheit, you simply multiply by $9/5$ and add 32:

$$°F = °C \times 9/5 + 32$$

Scientists often use absolute temperatures, measured in degrees Kelvin, named after Lord Kelvin, who invented a scale of temperature whose lowest value of zero corresponds to the point at which the molecules stop moving. This makes a lot of sense, given what temperature really is, and the scale is used by scientists worldwide. To convert from Celsius to Kelvin, you simply need to adjust for the difference between 0°C and the point at which the molecules stop moving. This turns out to be 273.14 degrees:

$$K = °C + 273.14$$

LAPSE RATE

Just eight minutes after leaving the Sun, light reaches the Earth. Some of this light is absorbed by the surface of the planet, where it warms the lower layers of the atmosphere, resulting in the temperature lapse rate: this is the rate at which temperature falls with height. In the atmosphere, this is around 0.6°C (1°F) per 100 m (300 ft) of altitude. This is why, as you climb to the top of a mountain, the air is cooler at the top than it is at the bottom and why the temperature outside when you take a plane flight is a chilly -40°C (-40°F).

MOLECULAR MOTION

The deeper we go into what temperature means, the more interesting it gets. The change in temperature of the atmosphere for a given energy input is called the heat capacity. This is the amount of energy needed to raise 1 kg (2¼ lb) of air by one degree. But what determines this value and hence the temperature of our atmosphere? It turns out that the more ways molecules can move, spin, or vibrate, which is called the number of degrees of freedom, the more energy we need to put in to raise the temperature. Since the molecules in the air are mostly oxygen and nitrogen, the atoms come in pairs, like tiny dumbbells connected in the middle with a flexible, springy bond. This means they can move in the three dimensions of space and rotate about two perpendicular axes (vibrations are also possible but not at the temperatures typical of the Earth's atmosphere). These five degrees of freedom dramatically raise the amount of heat energy needed to warm the atmosphere.

↓ Energetic molecules have translational motion in the three directions of space: V_y, V_x, and V_z.

↓ Diatomic molecules with lots of energy can spin about their axes in two perpendicular directions (Y and Z).

TROPOSPHERE

The troposphere is the lowest layer of the atmosphere, where we live and breathe. The name is derived from the Greek word *tropos*, which means "turning." This immediately reveals the main feature of the troposphere: it is constantly overturning.

It is this overturning that creates much of our weather, and almost all clouds, cyclones, and anticyclonic weather systems are found in the troposphere. What drives the constant turbulence of the troposphere is the energy from the Sun, which mostly passes through the atmosphere but is then absorbed at the surface of the Earth, causing heating of the lowest layers of the troposphere. This means that the warmest air sits in the lowest layers near the surface and colder air sits above. As a result, the troposphere is fundamentally unstable: it is prone to overturning.

CONVECTION

Pockets of warm air rising from the surface soon encounter lower and lower ambient air pressure as they rise through the troposphere, so they expand, cooling as they go, to conserve energy. When cold enough, water vapor in the air condenses out into droplets, forming clouds and releasing "latent" heat carried by the water in its vapor form. If the supply of vapor is sufficient, then rain will follow. It is this process of evaporation at the Earth's surface, and condensation above, which powers weather systems.

→ Deep convection in the troposphere transports water vapor upward where it condenses to form clouds and rain in a never-ending hydrological cycle between oceans, land, and atmosphere.

TROPICAL HEATING

The heat engines of the world's weather are in the tropics, where surface temperatures are highest, convection is the most intense, and vast quantities of energy are transported and released into the upper troposphere by condensing water vapor. The resulting disturbances in the atmosphere can then travel across the world, connecting tropical weather to cold snaps, floods, or heatwaves thousands of miles away. Like all physical connections between one place and another, this doesn't occur instantaneously. The message is carried by enormous planetary-scale waves called Rossby waves, which can travel across the globe in a matter of just a few days in the troposphere; akin to waves spreading from a pebble thrown into a pond.

STRATOSPHERE

Sitting atop the turbulence, clouds, rain, and storms of the weather below, is the stratosphere. The stratosphere is deep: it extends from the top of the troposphere (the tropopause) at 10–15 km (6–9 miles) above the surface, all the way up to around 50 km (30 miles) where the air is around a thousand times thinner (less dense).

STABILITY

The stratosphere was only discovered when intrepid balloonists dared to ascend through the troposphere, enduring extreme cold and using primitive breathing apparatus to cope with the low-density air. Some of them nearly died in the process, but they discovered that the cooling with height through the troposphere came to an end. Once they crossed the tropopause and entered the stratosphere, the temperature started to rise again. This means that, unlike the turbulent troposphere, warmer air rests on top of cooler air, with no tendency to overturn; the stratosphere is therefore stable and stratified, hence the name.

OZONE LAYER

Although it may seem remote and far from everyday experience, the stratosphere became a household word in the 1980s when it was discovered that the ozone layer, which sits in the lower stratosphere, was undergoing a rapid decline due to human-made chemicals: chlorofluorocarbons (CFCs).

The detection of the ozone hole is a remarkable tale of scientific discovery and a triumph of scientific enquiry over technological automation. In 1985, after checking and rechecking their data, Joe Farman and colleagues at the British Antarctic Survey in Cambridge eventually published their findings: the amount of ozone above Antarctica was in freefall. This took some courage since the NASA satellite monitoring the ozone layer suggested no such change and it was only later found that a hardwired error trap in the computer software was discounting observations of such low ozone levels because they were considered to be impossible at the time!

It was not long afterward that scientists unraveled the chemical mechanism and cemented the evidence that the actions of human beings could change the global atmosphere. Potentially serious consequences were envisaged for ecosystems and humans as increased ultraviolet began to stream through to the surface in the Austral (Southern Hemisphere) spring, when the hole was at its deepest. However, this tale ends on a positive note. In 1987, the Montreal Protocol was agreed by countries worldwide to inhibit the production and use of CFCs and related ozone-depleting chemicals. This was successfully implemented, with aerosol propellants and refrigerator coolants replaced across the globe. By the opening decades of the 21st century, ozone depletion had halted and the first reports of a recovery in the depth of the ozone hole were made.

THE WORLD AVOIDED . . .

Although it was not investigated at the time, the CFCs causing ozone depletion were also intensely powerful greenhouse gases. The evidence for climate change from carbon dioxide emissions was still being debated at the time and the problems from global warming were far less widely recognized than they are today. However, it's recently been shown that had we ignored the Montreal Protocol to reduce CFCs, and had their emissions continued unabated, then the additional climate warming effects today would have seriously compounded the issues of climate change we now face.

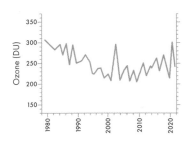

← Ozone amounts over Antarctica (in Dobson Units: 100 DU = 1 mm of gas at Earth's surface) between 1980 and 2020, showing the development and stabilization of the spring ozone hole.

MESOSPHERE

As we ascend through the atmosphere, the pressure reduces and the air becomes less and less dense. The last layer of what we normally think of as the atmosphere is the mesosphere. We are now far above the surface of the Earth, at 50–85 km (30–50 miles) altitude. The curvature of the Earth and the darkness of space beyond are now apparent.

The mesosphere is less well understood than the lower layers of the atmosphere, but it's home to some surprising and notable phenomena. Like the troposphere, the temperatures increase with height and there is lots of turbulence as disturbances from below stir the thin air. The world's highest clouds are found here too: so-called "noctilucent" clouds, which reflect sunlight from over the horizon after sunset. Bizarrely, only the summer mesosphere hosts the super-low temperatures needed for noctilucent clouds and it holds the record for the coldest place in the atmosphere at around -120°C (-184°F).

↓ Layers of the atmosphere showing the mesosphere 50–85 km (30–50 miles) altitude and the

circulation from the summer to winter poles (P) in the mesosphere. The Equator is marked (E).

→ Noctilucent clouds in the mesosphere are visible in summer just after sunset.

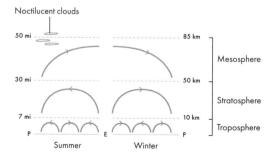

Noctilucent clouds

50 mi	85 km	Mesosphere
30 mi	50 km	Stratosphere
7 mi	10 km	Troposphere

P   E   P

Summer   Winter

GOING AROUND IN CIRCLES

In some situations, a change in pressure results in a change in the wind, as air rushes toward lower pressure and creates a sudden gust. For example, a sudden drop in pressure when your front door is opened can slam the back door shut as the air rushes through due to the drop in pressure.

The atmosphere, however, is more subtle. Unless you're in the deepest tropics (within just a few degrees of the Equator), winds do not flow toward low pressure and instead flow *around* pressure lows and pressure highs. This occurs because there are other forces at work and those forces arise because the Earth is rotating.

CORIOLIS FORCE

The fact that we measure everything in everyday life relative to the surface of the Earth means we are not in the unmoving, flat frame of reference that we think we are. In fact, we measure everything relative to the rotating frame of reference at our position on the Earth's surface. The effect of this rotation is small for things like driving in your car, but on large scales it needs to be taken into account. If you want to precisely calculate the motion of artillery projectiles, you must allow for this rotational effect. On the even larger planetary scales of weather systems, the effect can be treated as a dominant additional force: the "Coriolis force," named after French scientist Gaspard-Gustave de Coriolis (1792–1843), who first formulated it mathematically.

LOWS AND HIGHS

The Coriolis force is needed to explain the basic features of our low and high pressure weather systems. In the Northern Hemisphere the Coriolis force always acts to the right of the wind, while in the Southern Hemisphere it acts to the left. In a cyclone, the pressure force is toward the low pressure in the center; to balance this, the Coriolis force must act outward. Cyclones therefore circulate counterclockwise in the Northern Hemisphere and clockwise in the Southern Hemisphere. Given the Coriolis force acts to the right of the wind in the Northern Hemisphere and to the left in the Southern Hemisphere, it follows that there is no Coriolis force on the Equator. This is why hurricanes cannot form exactly on the Equator.

← Spinning cyclonic storms like these examples over the North Atlantic are a testament to the balance between pressure forces and the Coriolis force.

STORM TRACKS

The contrast between Earth's warm tropical climates and cool polar climates inevitably results in regions where the temperature lowers rapidly with latitude. This temperature gradient, coupled with the spin of the Earth, creates strong westerly winds.

The continental land masses and mountain ranges disrupt these winds, and the net result is a strong jet stream over the Pacific and Atlantic oceans. These regions are fertile nurseries for growing cyclonic windstorms, which feed off the wind shear (the increase in winds with height), and both the Pacific and Atlantic basins contain storm tracks.

CYCLONIC STORMS

As they track eastward, low-pressure storms deepen. The most intense storms can undergo "bomb" development when the central pressure drops on average by 1 millibar (mb) every hour over a day. These are the most intense storms and can reach the Eastern Atlantic with a drop of 50 mb or more in the central low pressure. Winds are strongest on the southern flank of cyclones and associated gusts cause most damage and loss of life during windstorms.

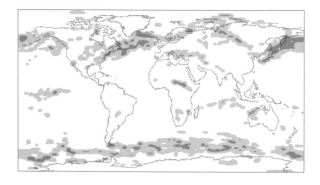

STORMY SEASONS

A lot of windstorms go through their life cycle over the oceans and their most devastating impacts are avoided. For example, after crossing the Atlantic, many storms pass between Scotland and Iceland, avoiding major land areas. However, if they follow a more southward track, they collide with densely populated areas in northwest Europe. One of the worst recorded storms in modern records was Storm Lothar, which hit Europe on Christmas Day in 1999 and caused multiple fatalities, widespread power outages, and billions of Euros of damage. Between December 2013 and February 2014, a whole series of Atlantic storms dumped record rainfall over the United Kingdom and caused widespread flooding for weeks on end.

The quick succession of storm after storm is thought to be a result of storm "clustering," where the leading storm creates ideal conditions in its wake for the formation of another. This behavior was repeated between December 2019 and February 2020 when a series of storms, including Storm Dennis with low pressure of 920 mb, one of the deepest centers ever recorded, again caused widespread havoc. On this occasion, in addition to storm clustering, there were remote influences from far away in the tropics that affected the jet stream and energized the storm track—scientists are continuing to research the causes of these intense and damaging weather seasons.

FUTURE STORMS

What will happen to the storm tracks in the future? To answer this question, we have the past weather records and computer models of the weather to help. These both show signs of the jet streams getting stronger and slowly migrating poleward. The storm tracks follow suit and are predicted to intensify and focus over northwest Europe.

← Storm tracks are found in the extratropics, where strong gradients of temperature and increasing winds with height provide the perfect breeding ground for cyclones.

IT'S ALL ABOUT WHICH WAY
THE WIND BLOWS

When we see a sudden change in the weather—for example, when the temperature suddenly drops during the winter—it is almost always associated with a change in the direction of the wind.

The idea that cold or mild air is carried along (advected) by the wind is a key principle of meteorology and provides a simple insight into why the weather one year differs so dramatically from the next. In general, advection toward the Equator brings cooler air masses and a drop in temperature and if air is exchanged with the colder polar regions it can lead to dramatic weather events.

COLD EUROPEAN EASTERLIES

During the winter of 2010, weather forecasters had long been warning of a cold start to the winter in Europe. In the event, the usual westerly winds from the Atlantic were replaced by easterlies, and instead of the usual flow of mild wet air from the Atlantic, easterly winds advected cold continental air from Siberia over western Europe. The result was one of the coldest Decembers in recent decades across northern Europe, with intense snowfall and freezing temperatures for weeks.

FUTURE COLD SNAPS?

How can intense cold snaps occur if climate change is meant to be warming the planet? After all, climate change has been occurring for many decades in the run up to these cold snaps. The answer to this apparent conundrum is that the same events would have been even colder had climate change not already warmed the air, and the air in the Arctic in particular, above the norm. For example, the chances of historic cold freezes over Europe, such as the United Kingdom winter of 1963, are now so unlikely that we will probably never see a similar event again.

COLD NORTH AMERICAN NORTHERLIES

Similar changes in winds affect regions worldwide. In February 2021, the whole of the North American continent was engulfed by Arctic air that flowed south over Canada and the United States. This again came from a simple change in the wind direction as the usual westerly winds buckled and dipped southward. Record low temperatures were recorded in some states, including Texas and Oklahoma, where disruption to energy distribution systems left millions of people without power and led to hundreds of deaths.

MONSOON WINDS

Changes in winds also deliver changes in rainfall and a switch to warm, moist air from maritime sources can herald sudden heavy rainfall. The most extreme examples of such flows are the monsoons of Asia, Africa, and Central America. The switch from dry to wet conditions in India is marked by a shift in the winds over the Arabian Sea to southwesterlies, bringing warm, moisture-laden air to India. This is such a dramatic switch that the monsoon onset date is set from historical records as June 1 and progressively later dates northward through the country. Any deviation toward an early or late onset of the monsoon is monitored and predicted each year with keen interest by meteorologists, since this switch in the winds marks the start of the four-month rainy season during which India receives most of its rainfall.

→ Typical onset dates for the Indian monsoon as it migrates poleward: (A) May 20, (B) May 25, (C) June 1, (D) June 5, (E) June 10, (F) June 15, Delhi June 29, (G) July 1, (H) July 15.

HADLEY CELLS

On average, the surface of the Earth receives much more energy from the Sun near the Equator than at the poles. Although the Equator and poles are both a similar distance from the Sun, the tropics are directly facing the Sun's rays and so absorb lots of energy, raising the tropical temperatures near the surface.

This warm humid air rises at the Equator, expanding and cooling so that the water it contains condenses. This results in heavy rainfall in a band that extends around the tropics, which is called the Intertropical Convergence Zone to reflect the fact that it is fed by air from both the north and south. All of this rising air then has to go somewhere; some of it descends locally but the rest spreads north and south, carrying lots of heat away from the deep tropics and into the Northern and Southern hemispheres. These are the Hadley Cells.

DESERTS

The air cools and starts to sink on its journey out of the deep tropics and eventually descends to the surface—where it is relatively dry with little rainfall. These regions contain high pressure and the deserts in a band around 20–30 degrees north and south of the Equator. Near the surface, the air then returns Equator-ward, slowing as it goes, due to surface friction, but this time the Earth's rotation means that the air is spinning in ever-increasing circles, so it has to slow down to conserve its spin about the Earth's axis and starts to flow from the East. These easterlies are the trade winds and were used for centuries by ships to cross ocean basins, carrying cargo and connecting one continent to another.

→ Hot deserts don't just occur anywhere. They are found in a band around the tropics where the Hadley Circulation descends and suppresses rainfall.

GEORGE HADLEY

The other important ingredient in the Hadley Cell is Earth's rotation. This was first realized by George Hadley (1685–1768) in 1735, hence the name. Due to Earth's daily rotation, everything at the Equator moves incredibly quickly—at around 1,600 km/h (1,000 mph), or 450 meters per second (m/s)! As the air moves away from the Equator north and south into the Hadley Cells, it is spinning in ever-decreasing circles around the Earth's axis. This means that it has to move faster and faster and a strong westerly jet stream is formed at around 30 degrees latitude in each hemisphere.

SEA BREEZES

O n summer days, the surface of the land heats up more rapidly than the ocean, since it is less able to cool by evaporating water. This creates a land-sea temperature contrast and lower pressure over the land than the ocean.

On these smaller scales, the Coriolis force (see Chapter 1, page 23) is a small factor and so an onshore "sea breeze" develops directly toward the low pressure. It can penetrate many miles inland and occur at the same time each day. Night-time cooling leads to the opposite effect and a relatively rapid cooling of the land surface produces an offshore or "land breeze." This means that the best time for calm conditions is often in the early evening.

The sea breeze is a mini weather front with convection and sometimes thunderstorms in a clear band running parallel to the coast. Sea breezes can also form over large lakes, but one of the more spectacular examples of sea breezes results in the "Morning Glory" roll clouds of northern Australia.

↓ During the day the land heats more quickly than the ocean, resulting in temperature and pressure differences that drive an onshore breeze.

↘ During the night, the reverse is true: the land cools faster than the ocean and the pressure and temperature difference drives an offshore breeze.

→ Morning Glory clouds are a dramatic example of sea breeze effects and the extended roll clouds that result can be seen appearing regularly in the morning at favorable locations.

Day

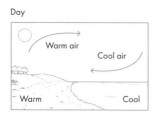

Night

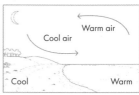

KATABATIC AND ANABATIC WINDS

Katabatic and anabatic winds are small-scale regional winds that occur on mountain ridges and plateaus. Katabatic winds flow downward and anabatic winds flow upward.

KATABATIC DOWNSLOPE WINDS

Katabatic winds occur when the land surface at high altitude is preferentially cooled by radiation, forming cold, dense air that then flows as a density current down the sides of mountain ridges into regions of lower density in the valleys below. These downslope winds can reach high speeds, over 100 mph (160 km/h) in some cases.

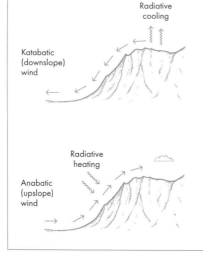

← Katabatic winds form when radiative cooling on mountains creates a density current that flows downhill, accelerating as it descends.

↙ Anabatic winds form when mountain slopes are heated by the Sun, causing rising motion and an upslope wind.

→ Antarctica hosts some of the strongest katabatic winds on the planet. Its cold climate and steep mountains are perfect for the formation of strong downslope winds.

Katabatic winds are common in Antarctica and are often cold, but as the air flows to lower altitude they are compressed by the increasing ambient pressure, thereby warming and, in some cases, such as in Santa Ana in California, becoming hot and dry.

ANABATIC UPSLOPE WINDS

Anabatic winds are the opposite and occur when preferential warming of the upper slopes of mountains reduces the air density, which leads to low pressure on the slope and convergence of air toward the slope, causing rising motion and an upslope wind. These winds occur mainly in the daytime and in summer, when they present good conditions for hang-gliders and small aircraft, but they can be much deeper and more turbulent than katabatic winds because the rising motion can easily form clouds, rain, and sometimes storms.

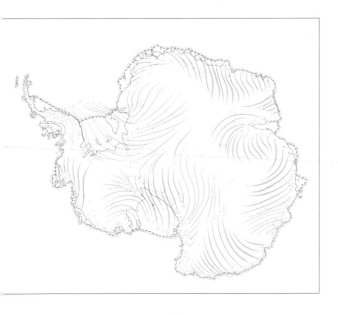

CLOUDS

As air rises, it experiences lower pressure and expands and cools. Eventually, the air is so cold that any water vapor it is carrying condenses out into tiny droplets and a cloud is formed. However, each droplet needs a tiny particle to seed its formation and if the air is completely clean, it can be supercooled well below the usual temperature before condensation occurs. The tiny nuclei needed for cloud formation often come from dust or salt particles over the ocean.

Clouds come in an infinite variety of shapes and sizes. High clouds start with the prefix "cirro," low clouds with "strato," and mid-level clouds with "alto." There are lots of subtypes, organized like Linnaeus's double-barreled names for biological organisms. So, for example, Stratocumulus lenticularis are stratocumulus clouds but in the shape of a lens. This particular type of cloud has such a smooth and unusual shape that they have even been mistaken for UFOs.

↓ When the Sun warms the surface of the Earth, pockets of warm air expand, rising due to buoyancy. The air expands and cools and water vapor condenses to form cloud.

↓ As air rises it cools due to expansion (D) until the lapse rate temperature is equal to the dew point temperature (C) and condensation occurs (B), forming the cloud base (A).

→ Clouds are classified into a number of types, depending on altitude and structure. The deepest cumulonimbus clouds can extend through the whole troposphere and create intense thunderstorms. (A) Cumulonimbus, (B) Cirrocumulus, (C) Cirrus, (D) Cirrostratus, (E) Altostratus, (F) Altocumulus, (G) Nimbostratus, (H) Stratocumulus, (I) Cumulus, (J) Stratus.

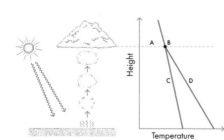

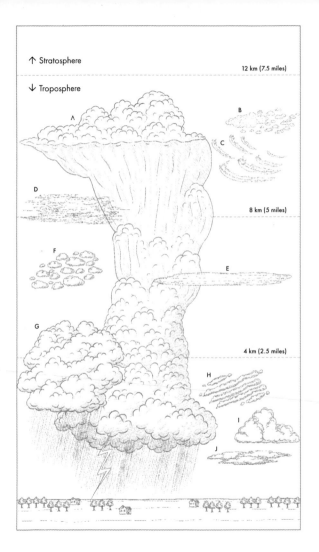

Stratosphere

12 km (7.5 miles)

Troposphere

A

B

C

D

8 km (5 miles)

F

E

G

4 km (2.5 miles)

H

I

J

RAIN

Rain forms when water vapor condenses out of the air as it is cooled, typically during ascent as the air rises into lower pressure, expanding and cooling. Water molecules find it easiest to condense out of the gas phase and into liquid water if there is something to stick to, and this can be just about any small airborne particle—for example, a speck of dust or tiny grain of sea salt. As more molecules condense, a raindrop is formed. Starting as microscopic drops in clouds, these coalesce and grow, then start to sediment out. Raindrops are typically a fraction of an inch (a few mm) in diameter since larger drops become distorted and split due to air resistance.

HYDROLOGICAL CYCLE

The process of rainfall forms a key step in the hydrological cycle, transporting evaporated water over the land so it can fall as rain and return to the ocean. Around 90 percent of rainfall originates from evaporation from the surface of water bodies like the ocean and most of the remaining 10 percent comes from transpiration by plants, where capillary action draws water up from the ground and through plant stems before releasing it into the atmosphere by evaporation from leaves. There is a broad range of times for water molecules to go around the hydrological cycle, but once in the deep ocean, they can spend thousands of years before re-entering the atmosphere.

Condensation to make rain releases heat and this warms the air, driving further ascent and condensation. This means that rainfall is to some degree self-sustaining and in the deep tropics, where there is plenty of moisture and plenty of warmth, we see the deepest rain clouds. They stretch all the way through the troposphere and try to overshoot into the stratosphere, where their ascent is curtailed by the highly stable stratospheric air. The energy released by tropical convection can be thought of as "weather fuel." It drives intense convergence of the air near the surface and divergence of air aloft. As the divergent flows approach regions of strong wind, they cause meanders in the jet streams that can travel across the globe as gigantic planetary-scale waves (see Chapter 11, page 17). If the tropical rainfall

GLOBAL RAINFALL

There is a global distribution to the Earth's rainfall, with high rainfall in a band near the Equator. This band of high rainfall, called the Intertropical Convergence Zone, coincides with the ascending part of the Hadley Circulation (see Chapter 2, page 28), and through the year it rocks back and forth across the Equator, following the Sun into the summer hemisphere. The highest average rainfall is found in South America, where some locations have annual rainfall over 10 m (30 ft) per year. The map below shows the distribution of rainfall (mm/day) in Northern Hemisphere winter, showing the intense rainfall bands of the tropics and the extratropical storm tracks.

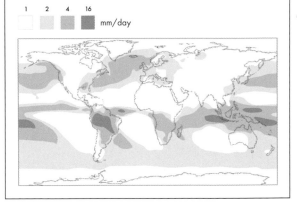

at their origin persists, these waves can lead to long periods of wet or dry, blocked weather patterns in the midlatitudes.

Earth's ecosystems rely on rainfall and scientists agree that without liquid water, it is hard to see how life can be sustained. Biodiversity across the planet is highly correlated with rainfall amounts and it is in the high rainfall regions of the tropics that we find rainforests and the highest biodiversity.

RAINBOWS

Rainbows are among the most stunning and familiar of all of our weather phenomena. Counterintuitively, they require both rainfall and sunshine to occur at the same time, which is why they are only seen occasionally.

The rainbow was explained centuries ago by the physicist Sir Isaac Newton (1642–1727) during his studies of optics and it arises from the process of refraction. The light from the Sun is not only made of the white color that we are accustomed to; it is, in fact, made up of light with a whole spectrum of colors. Each component color corresponds to a different frequency and a different wavelength, and

this means that as it passes through the surface of a falling raindrop, it is deflected (refracted) to a different degree. After entering the raindrop, some of the light is then reflected from the back surface of the drop to re-emerge through the front, where it undergoes a second refraction. So, if you stand with your back to the Sun and face the rain, you see the rainbow of colors.

RAINBOW ANGLES

The angle of the rainbow is always the same and is close to 42 degrees. The reason the rainbow is an arc is simply that the ground gets in the way, so if you are lucky enough to spot a rainbow while flying at altitude you can sometimes make out the whole circle. In very bright examples of rainbows, it is also sometimes possible to make out a fainter but discernible second bow, concentric but outside the main bow. This is caused by light that undergoes two reflections in the raindrops and the colors switch order compared to the first bow.

WAVELENGTH AND COLOR

Splitting the white light from the Sun into its component colors produces a spectrum that ranges from red, which has the longest wavelength, to violet with the shortest wavelength. It is the shorter wavelengths that are refracted most as the light enters and leaves the raindrop, and this is why violet is on the inside of the rainbow and red is on the outside.

← Rainbows result from the refraction of sunlight by raindrops, so you need simultaneous sunshine and rain to see a rainbow. Multiple bows like the one shown here correspond to multiple reflections of the light inside each water droplet.

HAIL

Hailstones are not stones at all of course, but compact, hard balls of ice that have accreted onto a small nucleus while being suspended high in the atmosphere before falling out and reaching the surface. If you slice a good-size hailstone in half, you can see the rings, much like tree rings, where layer after layer of ice has accreted onto the hailstone.

GLOBAL HOTSPOTS

Hailstorms can be very costly events and are one of the weather hazards that do the most damage to crops and property. The Australian hailstorm of April 1999 cost billions of dollars and was one of the costliest weather events in recent Australian history. Understandably, insurers are keen to know where the hotspots for hailstorms are around the world, but this is not always easy to map out because hailstorms can be isolated, small-scale events that often go unobserved. Nevertheless, we know of a number of hotspots for hailstorms. The mid-United States holds the record

---

HAILSTONE SUSPENSION

The conditions needed to produce a really intense hailstorm include lots of surface warmth, which leads to lots of rising motion, and lots of wind shear—the change in winds with height—which helps to spin up the storm. Intense rising motions are needed to support the hailstones in mid-air: a 1-cm ($1/2$-in) stone needs the air to be rising around 10 m (30 ft) every second to be supported, and this increases with the size of the hailstone. Keeping the hailstones above the melting level for long enough to grow is important because the formation of large hailstones requires the capture of as much freezing water and other hailstones as possible.

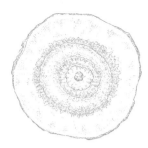

← Hailstones form as water accretes onto a central nucleus. As the hailstone enters different environments within the cloud, it accretes different-sized particles of water and ice, creating layers like those found in an onion.

for the most frequent hailstorms, where localities have 20 days of hail in a typical year, but there are also other hail hotspots in equatorial Africa, Southeast Asia, Argentina, southern Africa, and the Mediterranean.

GIANT HAILSTONES

Hail is usually around 5 mm (¼ in) in diameter and small hail often melts as it falls, descending below the melting level into the warmer atmosphere below and turning back into rain. However, the biggest hailstones can be bigger than your fist and pack a massive punch as they hurtle to the surface because for every doubling in size of the hailstone its energy goes up by 16 times!

FUTURE HAILSTORMS

What will happen to hailstorms in the future as the climate warms? Will the warmer atmosphere melt hailstones so that they become less frequent? Or will hailstorms become more intense and cause more damage? The answer to this question is still debated and there are competing effects, but it seems that the extra warmth near the surface means that small hailstones will be less common as they will more likely melt. However, the warmth will also give rise to more intense rising motion and hailstorms, and the raising of the melting level due to the warming of the air in the lower atmosphere means that only the large hailstones will survive to the surface before melting. Overall, this means we can probably expect record-breaking hailstones in future.

SNOW

Snow is the remarkable sub-zero alternative to rain if there's enough moisture around. Snowflakes are made of water crystals and most are a conglomerate of different individual crystals. There is lots of air in snow and little water, so 10 cm (4 in) of snow on the ground corresponds to just 1 cm (1/2 in) of rain.

6-FOLD SYMMETRY

Why do all snowflakes have six points? Water molecules have two hydrogen atoms and one oxygen atom in a V-shape with an angle of 105 degrees. This isn't too far from 120 degrees, so, as water molecules freeze, they produce regular hexagons, but it's impossible to predict the shape of a snowflake in advance because new molecules joining the crystal are affected by the slightest change.

Some of the deepest snowfalls in the world are found in Japan, where 30 ft (10 m) of snow is common during winter northerlies. So perhaps it is no coincidence that 20th-century Japanese researcher Ukichiro Nakaya was the first to show that snowflakes always come as plates or columns, depending on humidity and temperature.

↓ Water molecules have a bent structure with the two hydrogen atoms of H_2O forming an angle of 104.5 degrees.

↓ Water molecules bond in hexagonal, repeating structures, which have 120-degree bond angles.

→ Microscopic 6-fold symmetry is expressed in the macroscopic shapes of snow and ice crystals. An infinite variety of shapes is possible and the exact type depends on details of the ambient temperature, pressure, and humidity.

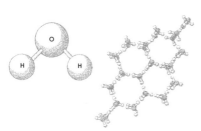

THUNDER AND LIGHTNING

Lightning is an electrical discharge that occurs between thunderstorms and the ground or, more commonly, within clouds. The lightning "bolt" is a gigantic spark that allows an electric current to flow and releases large amounts of heat and light. The air gets so hot during lightning that it expands faster than the speed of sound, and this creates a shock wave that travels through the atmosphere as the booming sound we call thunder.

The flash of lightning and boom of thunder also give rise to the common method of determining how far away a thunderstorm is. Compared to the sound, the light travels almost instantaneously, so the light arrives at your eye almost straightaway, while thunder arrives after the sound has had time to travel. The speed of sound is around 300 m/s (300 yards per second), so if you count the seconds between flash and boom and multiply by 300, you get the distance to the storm, and if the time gets shorter each time, then the storm really is getting closer.

ELECTRIC CHARGE

We still don't understand all of the microscopic processes at work in a thunderstorm but rapid rising motion causes friction and collisions between lighter ice crystals and heavier falling graupel (granular pellets). This rubs off electrons onto the graupel, leaving the positively charged ice crystals to rise to the top. A few hours later, the charge difference between the positive cloud top and its negative base can result in a voltage difference within the cloud (or with the ground) of many millions of volts. Such high voltage differences can lead to electrical breakdown and a sudden drop in resistance as electrons flow through narrow forked channels called leaders, rushing to the Earth in a tiny fraction of a second as lightning strikes.

LEADERS AND STREAMERS

As a thunderstorm passes overhead, the negatively charged cloud base repels electrons in the atoms at ground level and the surface becomes positively charged. If the effects are strong enough, positive charges start to make the air glow and flow upward in a streamer as they try to connect with the negative leaders coming down from the cloud base. These upward streamers are most often seen reaching out from high points like tall buildings just before a strike.

↓ Lightning is a gigantic spark that resets the slow buildup of electrical energy when the voltage difference is enough to overcome atmospheric air resistance. Thunder results from the supersonic velocity of the heated air.

TORNADOES

U sually a few hundred meters in diameter, tornadoes are among the smallest extreme weather systems. Tornadoes are cyclones and have low pressure at the center, but unlike hurricanes or midlatitude windstorms, the Coriolis force is negligible in these small vortices. Instead, the push of the air toward the low pressure in the tornado is balanced by the centrifugal force due to its rapid rotation. The Fujita scale is used to measure tornado strength: the strongest category (F5) tornadoes have winds over 415 km/h (260 mph).

Tornadoes can travel a few miles before they "rope out," narrowing and twisting as they break apart. They're small compared to the resolution of typical weather forecasts, so meteorologists concentrate on predicting the strongly convecting, sheared wind regions that provide good spawning grounds for tornadoes. Tornado alley in the mid-United States is famous, but they are found in many regions of the world—when they form over the ocean they are known as waterspouts.

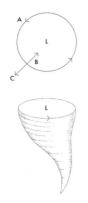

↙ Forces in a tornado. The pressure gradient force inward (B) is balanced by centrifugal force outward (C). This results in fast cyclostrophic winds (A) in small-scale, rapidly rotating vortices such as tornadoes.

↙ The low pressure inside a funnel cloud (L) causes the water vapor in the air inside to condense, which is why we see the funnel. It only becomes a tornado if the funnel reaches the ground.

→ Tornadoes are small but devastating columns of rapidly rotating air that form at the base of cumulonimbus clouds in severe convective storms. A single storm can produce multiple tornadoes and a typical tornado will wreak havoc along a path of a few miles. Tornadoes almost always rotate cyclonically despite the negligible effects of the Coriolis force on their small scale since they take their spin from the parent storm.

HURRICANES AND TYPHOONS

Hurricanes and typhoons are the classic spiral storms you see in satellite photos of the Earth from space. Hurricanes and typhoons are actually the same phenomenon, it's just that typhoon is the name for a Pacific storm and hurricane is the name for an Atlantic storm—they are both examples of tropical cyclones.

FORMATION

Hurricanes and typhoons form over the warm waters of the tropical Atlantic and the tropical Pacific. However, they need to feel the Earth's rotation and so they can't form right on the Equator. Instead, they usually develop between 5 and 20 degrees North. The storms develop from smaller convective weather systems; in the Atlantic case, these are often associated with wavelike disturbances traveling westward off the African continent.

Developing hurricanes and typhoons travel westward in the trade winds and strengthen as they extract energy from the warm ocean surface. Air spirals in toward the center of the storm near the surface and outward at the top of the storm. Intense convection forms in

---

CONDITIONS FOR TROPICAL CYCLONES

Hurricanes and typhoons only form when the sea surface is warm enough and typically above 26°C (79°F). They also require weak wind shear between the easterly trade winds near the surface and the westerly winds aloft. If these are too strong, they can shear out and disrupt the storm, preventing it growing. There are year-to-year changes that make some years active and some years weak for tropical cyclones. For example, although it delivers extreme weather in other forms across the globe, El Niño (see Chapter 5, page 60) tends to increase the wind shear in the Atlantic and reduce the intensity and number of hurricanes.

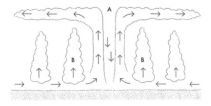

← The air flows into a hurricane near the surface, carrying warm moist air to power the storm. Cloud bands (B) spiral around the central eye (A) where descending air creates clear skies and a lull in the wind.

bands around the storm center and the condensation of moisture aloft acts like a heat engine, adding heat to the air and powering it toward ever-stronger winds.

All tropical cyclones have low pressure in the center and rotate counterclockwise in the Northern Hemisphere. Southern Hemisphere storms are less common and rotate clockwise. In a mature storm, a small region of descending air forms in the center. As the air descends, it is compressed and warms, so there is no condensation, cloud, or rainfall here and this eye of the storm is clear and quiescent. It is surrounded by a towering ring of deep storm clouds known as the eye wall, where some of the strongest winds (up to 320 km/h/ 200 mph) are found.

TROPICAL CYCLONE DISASTERS

Typhoons and hurricanes are the most energetic and dangerous storms on the planet. Typhoon Haiyan (aka Yolanda) tracked westward in the Pacific, growing to the highest category we have for tropical storms—Category 5 on the Saffir–Simpson scale—with the strongest winds ever observed in a tropical cyclone to that date. Haiyan made landfall in the West Pacific in early November 2013, where it devastated large areas of Southeast Asia. The Phillipines was among the worst countries affected and more than 6,000 people were killed by Haiyan.

Hurricane Katrina in August 2005 was the worst Atlantic hurricane disaster in modern times. It entered the Gulf of Mexico, where it was boosted by warm waters and made landfall over Louisiana and Mississippi. Heavy rainfall combined with a massive storm surge almost 10 m (30 ft) high flooded New Orleans for many weeks and Katrina claimed almost 2,000 lives.

EXTREME RAINFALL AND FLOODS

Inland flooding can have many exacerbating factors, ranging from changes in land use to changes in river courses, but the one driver that almost all inland flooding events have in common is extreme rainfall.

UNITED KINGDOM, 2007

Floods occur around the world and, in summer 2007, the Atlantic jet stream meandered southward of its usual position, bringing a persistent series of slow-moving cyclonic weather systems that hit northwest Europe and lingered over the United Kingdom. Much of England was affected and some regions received a month of rainfall in a single day, causing a whole series of floods. Towns were inundated with water and the iconic image of Tewkesbury Abbey, in Gloucestershire, under water for the first time in over two centuries, covered the front pages of newspapers. Hundreds of thousands were left without drinking water and flooding was also widespread in Wales and Scotland. These floods prompted a government review and new policies to tackle future flooding in the UK.

PAKISTAN, 2022

It had been some years since the summer of 2010, when Pakistan had last experienced really extreme prolonged summer rainfall, but in 2022 the summer rainfall record was set to be broken. A supercharged monsoon season, exacerbated by La Niña conditions (see Chapter 5, page 60) in the neighboring Pacific and climate change, sent low pressure systems extending west into Pakistan and excessive summer monsoon rains. The country received nearly three times its normal summer rainfall and the extra rainfall inundating Pakistan was more than five times the increase seen in a usual wet summer. After the Indus River burst its banks, more than 10 percent of the country was flooded, with millions left homeless and nearly 2,000 lives lost. The impacts of this devastating flood, the worst in many decades, were costly and longlasting.

COASTAL FLOODING

Other types of flooding can also occur and the most common of these is coastal flooding. Stormy weather conditions associated with deep, cyclonic low pressure systems can give rise to a storm surge. If this coincides with the timing of a high tide, then the temporary rise in sea level can top flood defences and give rise to coastal flooding at many places around the world where there are storms to deliver onshore winds.

↓ Monsoon rains occur every year in many tropical regions. Just a 10 percent increase in average rainfall can cause major flooding, with lives lost and devastating impacts on agriculture.

DROUGHTS

U nlike many weather and climate extremes, droughts start slowly. Persistent weather patterns can lead to a slowly accumulating rainfall deficit and, as the deficit builds, soil moisture declines and river and groundwater levels fall. The lack of soil moisture means that more of the energy arriving from the Sun goes into warming the land surface than evaporating water, and this exacerbates the high temperatures often associated with drought. Marking the exact start and end to a drought is difficult, but they can last for weeks, months, or even multiple years, as happened during the 1930s Dust Bowl (see Chapter 6, page 80).

SAHEL DROUGHT

One of the worst droughts in living memory occurred in the Sahel region of West Africa in the late 1970s and early 1980s when a decade-long drought devastated crops and caused widespread famine. At the time, there was concern that the region was turning into an extension of the nearby Sahara desert and that the main cause was land mismanagement. However, we now know that the Sahel drought was at least partly driven by slow variations in ocean temperatures and enhanced warming of the southern oceans relative to those in the north. This displaced tropical rainbands southward and away from the Sahel. Later in the 1980s, these ocean patterns changed and although individual years have experienced further droughts, Sahel rainfall has generally increased since the 1980s.

ENSO AND DROUGHT

Not all droughts have an obvious cause, but the El Niño–Southern Oscillation (ENSO)—see Chapter 5, page 60—is implicated in many large-scale droughts. Droughts are more likely in southern Africa, India, and the Amazon during El Niño, whereas California and East Africa are more prone to drought during La Niña. Some of these regions experience multiple years of drought when ENSO conditions persist; the wildfires of California and the persistent drought of East Africa between 2020 and 2022 have been connected to three consecutive years of La Niña conditions.

Just about any part of the world can experience drought, and droughts are predicted to increase in frequency and intensity due to climate change and global warming. It is now thought that drought affects more people than any other form of natural disaster.

← Prolonged lack of
rainfall drives drought,
heavily impacting
people, agriculture,
economies, and
ecosystems. Droughts
occur on all timescales,
lasting from weeks
to years.

EXTREME RAINFALL AND LANDSLIDES

Landslides are the mass movement of rock and earth downslope. They often occur suddenly and unexpectedly and they move at a rate that far exceeds any chance of escape for those below. Although landslides often have multiple causes, such as earthquakes and volcanoes, the major culprit by far is heavy rainfall.

SIERRA LEONE, 2017

Many landslides occur in remote regions and can be viewed as part of the natural weathering processes of the Earth's surface. However, others are much closer to home and create human disasters. One of the worst landslides of recent decades occurred in August 2017 in Freetown, the capital city of Sierra Leone on the west coast of Africa. The city had been subjected to days of heavy rainfall, which, coupled with destabilizing deforestation, led to the collapse of a large section of Sugar Loaf mountain, overlooking Freetown. The mudslides hit in the early hours of the morning and more than 1,000 lives were lost.

↓ Rain lubricates slip surfaces between rock layers, causing landslides. If it gets into sedimentary rocks with clear bedding, it can form a planar slip surface and lead to a translational landslide.

↓ If rainfall penetrates a curved, "spoon-shaped" slip surface, it leads to a rotational landslide, which tends to self-stabilize as material piles up at the base.

→ Persistent rainfall can destabilize hillsides by penetrating deep into soils and rocks. This creates a slip surface and leads to a catastrophic failure as the landslide occurs. Landslides are commonly "spoon-shaped," rotational type, or planar translational type, depending on the underlying geology.

WHAT'S POSSIBLE?

In this chapter we've looked at some of the most extreme weather and climate events seen in recent decades. In recent years, we have seen weather records smashed time and again. Heatwaves, drought, and rainfall records devastate crops, create destruction, and take lives.

CHAOS

We know that the atmosphere is "chaotic," meaning that only the statistics of weather are bounded by the current climate, and on individual days or weeks (see Chapter 8, page 100), individual events could easily evolve far differently due to the smallest of variations in the weather beforehand. In this case, we can ask: what's possible? How extreme could weather be? What's the worst-case scenario and could it be more extreme than has yet been seen?

COMPUTER SIMULATIONS

To answer this question, scientists working on weather and climate have used computer simulations of the atmosphere, altered by tiny but plausible amounts at the start of the simulation to generate a whole spectrum of ensuing weather patterns. Note that the computer models are based on the same equations that govern the atmosphere and while the exact weather patterns they produce are different in detail, they can be statistically indistinguishable from the real-world observations, and we can use the output of thousands of simulations with these models to explore what is possible and what is the worst case for extreme weather.

UNPRECEDENTED EXTREMES

The outcomes of these computer model experiments are a sobering call to be prepared for extreme weather. Almost without fail, they show physically plausible weather events more extreme than anything seen before. These include extreme heatwaves in China and the United Kingdom, as well as record-breaking temperatures that would be off the scale compared to current measurements. There are physically plausible rainfall levels that would smash the records and deliver unprecedented flooding. Indian monsoon droughts are simulated that are three times more intense than usual and well beyond even the most severe drought of the 20th century. These simulated events have not been seen so far, but the point here is that they are simulated by the same weather models we use to predict the weather every day and to make successful forecasts and simulate climate change. The models are tried and tested, and, while not perfect, in many senses they are indistinguishable from the real weather, so we should heed their warnings. These unprecedented events, while extreme and record-breaking, are possible now and, as far as we know, any one of them could occur in the coming year.

NORTH ATLANTIC OSCILLATION

For countries around the North Atlantic basin, winter conditions can be quite different from one year to the next. Some years are marked by very mild winters, while others are intensely cold. If we look at the differences between one winter and another, we often see a similar pattern: the North Atlantic Oscillation (NAO). The NAO is the single most important factor that determines the harshness of winter in North America, Europe, and North Africa.

↓ Snow in many parts of northern Europe and eastern USA is linked to the NAO. When the NAO is in its negative phase, the Atlantic jet stream weakens and cold air floods in.

EFFECTS

Sir Gilbert Walker, a British meteorologist, first defined and named the NAO after studying weather records in the 1920s. It has a particular pattern with low pressure over Iceland and high pressure over the Azores islands. When the NAO is positive, this pattern is strengthened and the pressure across a broad region centered over Iceland is lower than normal, while that in a broad region over the Azores is higher than normal. In this phase, the Atlantic jet stream has strengthened and moved northward and this means that northern Europe is milder than usual but also wetter and stormier as more Atlantic cyclones arrive there. In contrast, the Mediterranean and North Africa are cooler during this positive phase of the NAO. These effects also extend over the western side of the Atlantic basin, where the eastern United States is mild and eastern Canada is colder when the NAO is positive. Overall, this gives a four-fold pattern of temperature and it means that east coast United States temperatures vary with those in northern Europe.

CLIMATE VARIABILITY

The NAO has varied dramatically and caused some extreme winters in the past. During the winters of the 1960s, the NAO was often in its negative phase and northern Europe experienced some of its coldest winters of the 20th century. Then during the 1990s, the NAO flipped into its positive phase and a string of mild winters in northern Europe were marked by flooding and storm damage. These fluctuations continue and 2009/10 experienced the lowest NAO on record with intense cold and heavy snowfall across much of Europe, while the winter of 2019/20 had a positive NAO, with a battery of Atlantic storms and record late winter rainfall in the United Kingdom. Understanding and predicting these fluctuations is a hot topic with meteorologists.

EL NIÑO AND LA NIÑA

The largest, naturally occurring weather change from one year to the next is the El Niño–Southern Oscillation (ENSO). It comes in two phases, corresponding to warming (El Niño) or cooling (La Niña) of the tropical Pacific. Differences between La Niña and El Niño are just a few degrees of ocean temperature, but the impacts on world weather are far-reaching. Even global temperatures are affected and every degree of El Niño warming gives a temporary increase of about 0.1 degree in global temperature the following year.

ORIGINS

The El Niño and La Niña names are Spanish for "the boy" and "the girl," respectively, and they originate from the local inhabitants of South America who were all too familiar with the episodic changes in their climate. The name refers to the Christ child, as ENSO cycles tend to peak around Christmas. Rainfall and temperatures along the west coast of South America closely follow the ENSO cycle and fish catches rise during La Niña and fall during El Niño as the amount of nutrient-rich water is modulated.

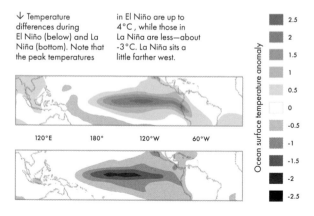

↓ Temperature differences during El Niño (below) and La Niña (bottom). Note that the peak temperatures in El Niño are up to 4°C, while those in La Niña are less—about -3°C. La Niña sits a little farther west.

Ocean surface temperature anomaly

2.5
2
1.5
1
0.5
0
-0.5
-1
-1.5
-2
-2.5

120°E   180°   120°W   60°W

NATURAL CYCLES

ENSO is coupled to the atmosphere and the ocean. As ocean temperatures cool along the equatorial Pacific during La Niña, the Pacific trade winds strengthen so that La Niña years are the best for trade ships to cross from South America to Asia. We know that ENSO cycles have been going on for many years, since Peruvian fishers have been following them closely to guide their fishing. This has now been extended to many thousands of years using coral records, which show periodic bleaching from the hot ocean conditions that come and go with ENSO.

CLIMATIC EFFECTS

The cycle usually ramps up in the summer and one of the first regions affected is India, where the monsoon rains often falter during El Niño, giving rise to droughts and reducing crop yields. ENSO cycles reach their peak at the end of the year and the eastward shift or rainfall during El Niño leaves behind drought and wildfires across West Pacific countries and northeastern Australia. The rainfall and rising air over the East Pacific is balanced by descent over northern parts of South America, and the Amazon rainforest also suffers from droughts and increased wildfires. Knock-on effects are felt as far away as southern Africa, which suffers from serious droughts during El Niño, and even northern Europe is affected, with colder, drier winters during El Niño and wetter, stormy winters during La Niña. Even after ENSO subsides in April, it leaves behind further weather changes in store; the summer monsoon in China is often supercharged following big El Niños, giving rise to flooding and destruction in the Yangtze River valley.

PREDICTABILITY

El Niño and La Niña are also among the most predictable phenomena in Earth's climate. Computer models, loaded with observations of what the ocean and atmosphere are doing today, can predict the cycle out to many months ahead and, in some cases, when a big El Niño is looming, forecasts can predict it more than a year ahead. This is crucial for some parts of the world, as ENSO cycles can drive droughts and floods across the whole globe.

MADDEN–JULIAN OSCILLATION

The Madden–Julian Oscillation (MJO) is the largest source of month-to-month changes in tropical weather. It was discovered recently, in the 1970s, as weather data gained greater coverage and revealed its prominent cycling, roughly every 40 days. There is a convective center to the MJO, with upward motion and heavy rainfall, balanced by neighboring descent and clear skies. However, the MJO is a traveling phenomenon and the whole pattern marches from west to east across the tropics at a speed of around 20 km/h (12 mph).

CONVECTIVE CORE

The convective center of the MJO is ripe for thunderstorms and even tropical cyclone development, so the risk of Pacific typhoons and Atlantic hurricanes varies with the MJO cycle. It also reaches into the extratropics, where it changes North Pacific winds, and as its center of high rainfall moves to the central Pacific, there is often heavy rainfall over western North America. The MJO even affects European weather, where it can help drive intense winter cold snaps.

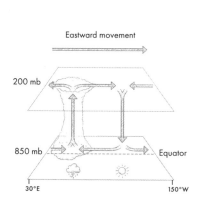

Eastward movement

200 mb

850 mb — Equator

30°E   150°W

↙ The MJO contains a region of strong convection with rising air that flows outward in the upper troposphere as it moves slowly eastward. It affects typhoons and weather out into the extratropics.

→ The MJO dominates tropical weather variations from month to month, from the Indian Ocean to the Pacific. It moves eastward over 40–60 days. Intense periods of MJO activity can trigger El Niño events in the East Pacific.

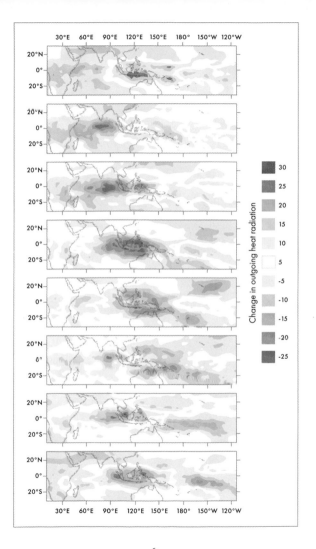

Change in outgoing heat radiation

SUDDEN STRATOSPHERIC WARMING

T he winter stratosphere is home to the largest cyclone on the planet—the stratospheric polar vortex. With low pressure over the pole, temperatures many tens of degrees below zero, and strong westerly winds circling the planet, this planetary-scale winter vortex is ripe for one of the most dramatic weather events on Earth.

DISCOVERY

This was first noticed by Richard Scherhag, a German meteorologist who had started to compile regular balloon-borne observations of the stratosphere. In early 1952 he published the results of his observations, which showed a startling change. The temperatures over the Arctic in the winter stratosphere showed a sudden and unexpected increase of many tens of degrees in just a few days. Nowhere else in the atmosphere had such a dramatic warming ever been observed and it prompted Scherhag to call it "explosive." Scherhag had witnessed the first recorded sudden stratospheric warming.

WEATHER IMPACT

After regular observations and monitoring became available and there were enough events to make a good-size sample, it was noticed that the period following a sudden stratospheric warming was often cold over northern Europe and the eastern United States. It turns out that the warming in the lower stratosphere, accompanied by weak or easterly winds as the vortex collapses, has an influence on the weather below and even at the surface. In the last few years, after intense cold and snow events following sudden stratospheric warmings like the one in early 2018, this high-altitude phenomenon is now one of the regular tools that long-range forecasters use to predict the weather.

MECHANISM

The air in the stratosphere is tenuous and, compared to the much higher density troposphere, changes in stratospheric winds can be rapid and dramatic. It was not until 20 years later, in the 1970s, and through the observation of a number of other sudden warming events, that Japanese meteorologist Taroh Matsuno presented a solution to what caused the rapid warming. In fact, there is no heating going on at all during a sudden stratospheric warming. Instead, the winds around the polar vortex are decelerated by enormous atmospheric waves that break on the edge of the vortex, causing the air to fall into the polar vortex, where it is compressed and therefore warms in a manner similar to the warming of the air in a bicycle pump as it is compressed to inflate a tire.

FREQUENCY

Sudden stratospheric warmings occur relatively frequently in the Northern Hemisphere and on average we see them in five or six winters per decade. However, this varies a lot and in the 1990s there was a run of almost ten years with no sudden stratospheric warmings at all. Southern Hemisphere events are much rarer and the first event was not recorded until 2002, when it took meteorologists by surprise. Since then, computer model experiments have been used to simulate thousands of virtual winters and estimate that we should see sudden warming events in the Southern Hemisphere roughly every 25 years on average.

→ The usual westerly winds in the polar vortex (left) collapse during a sudden stratospheric warming and air floods into the Arctic (right), compressing and warming as it sinks through the stratosphere (A) and leading to high pressure in the troposphere (B).

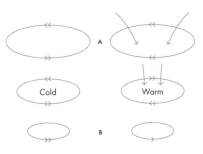

QUASI–BIENNIAL OSCILLATION

The regularity of the Quasi–Biennial Oscillation (QBO) is utterly remarkable. It consists of alternating westerly and easterly wind cycles that migrate slowly downward through the stratosphere, terminating at the boundary with the troposphere. These winds go right around the Earth in a belt in the QBO and so if you chart them by height and time over a number of years, you get a striking series of stripes as the oscillation flips from one state to the other.

↓ After the seasonal cycle, the QBO is the most regular feature of our atmosphere. It consists of oscillations between westerly (yellow) and easterly (blue) winds circling around the Equator at high altitude and propagating down through the atmosphere. The striking cycles of the QBO can now be reproduced in our computer models of the weather.

→ The QBO is driven by atmospheric waves that propagate up from the troposphere and approach the QBO from below, where they break like waves on a beach and drive the oscillation.

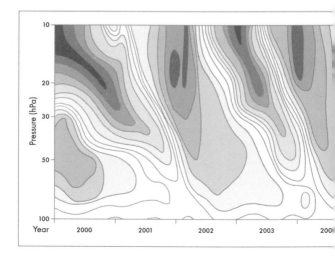

66

PREDICTABILITY

The winds are a little stronger and migrate downward a little more slowly in the easterly phase, but the QBO is the most predictable variation in the atmosphere and skillful forecasts are possible even years ahead. The QBO is all the more remarkable because its regular cycles are not forced by the Sun or other external driver. Instead, it arises spontaneously within the atmosphere in a process akin to longshore drift in the ocean.

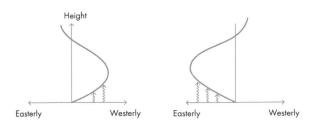

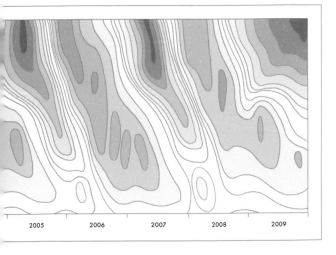

PACIFIC AND ATLANTIC VARIABILITY

Weather and climate fluctuations from day to day are usually greater than those from month to month, which in turn usually exceed those from year to year and so on. However, the decade-to-decade variations of the surface Atlantic Ocean are an exception. So-called Atlantic Multidecadal Variability (AMV) shows larger variations on decadal timescales than are seen from year to year. This unusual phenomenon fluctuates with changes in the poleward flow in the Atlantic Ocean that push more/less warm water poleward and create warm/cool Atlantic ocean conditions in the positive/negative phase. AMV stimulates droughts in North America and Africa and creates slow, decade-long variations in European summer weather and the strength of the Atlantic hurricane season.

PACIFIC CLIMATE

Although it is different in character, there is also prominent decadal variability in the Pacific Ocean. So called Pacific Decadal Variability (PDV) is closely related to multiyear variations in the cycling between El Niño and La Niña, although some climate scientists argue that it is independent. In the positive phase, the PDV displays warm conditions in a broad swath across the tropical Pacific. PDV also has far-reaching consequences, with regional climate effects similar to El Niño and La Niña and recent evidence that it was a key player in the so-called "global warming slowdown" of the early 21st century when it helped to temporarily slow the rate of global warming. PDV also changes the frequency of cold snaps over China from one decade to the next and it has a well-known impact on salmon abundance and fish catches off the coast of Alaska, which are enhanced when the PDV is in its warm phase.

ATLANTIC-PACIFIC LINK

Although they are separated by the Americas, Atlantic and Pacific decadal variations are now known to be connected. Analysis of historical climate records and experiments with computer models of the climate show that when the Atlantic warms it tends to trigger Pacific cooling. Although the changes are just a fraction of a degree, if the two ocean basins change in concert, they can trigger extreme events like the 1930s Dust Bowl (see Chapter 6, page 80).

← Decade-to-decade variations in summer rainfall in Europe are linked to Atlantic Multidecadal Variability.

D-DAY LANDINGS

In the latter part of the Second World War, the Allied forces secretly planned to invade occupied Normandy in France with thousands of ships and planes. The weather was crucial to the successful execution of this operation and minimal cloud cover with high visibility was needed. Other factors included the need for a low tide and plenty of moonlight once troops were deployed. These lunar conditions limited the opportunities to just a few days each month and it was important the operation was carried out early enough to stop German Field Marshall Rommel bolstering his Atlantic defence lines. Operation Overlord was set for the June 5, 1944, but only if the weather met the necessary conditions.

CRUCIAL FORECASTS

The need for accurate weather forecasts was soon realized by the Allied commanders; to this end, British Group Captain James Stagg was appointed Chief Meteorologist, responsible for disseminating forecasts to all the Allied forces. In the run up to the invasion, Stagg scrutinized the latest weather charts using Allied weather observations but also those made by German observers because, unbeknown to them, codebreakers at Bletchley Park, in England, had cracked the German Enigma code and were able to intercept and read the latest weather observations being transmitted from occupied Europe.

INFORMED DECISIONS

In early June, Stagg was concerned about a cyclonic low pressure system just to the north and he forecast strong winds and rough seas for June 5 that would make small boat crossings and low flights across the Channel impossible. Instead, he spotted a short weather window on June 6, when high pressure would provide the right conditions for the invasion.

General Dwight D. Eisenhower, supreme commander of the Allied forces, took Stagg's advice and delayed the invasion. This proved crucial and the landings on June 6 were met with high pressure and light winds. The only other plausible days in June were subject to high winds and poor conditions, and US President Truman later remarked that June 6 was "probably the only day during the month of June on which the operations could have been launched."

← Accurate weather forecasting was crucial to the timing of the D-Day landings, which serves as a dramatic reminder of how weather forecasts can save lives.

PESTS AND PLAGUES

The majority of insect species are dependent on the weather at most stages of their life cycle and these relationships occasionally lead to massive positive feedback on insect populations. For example, in 1976, the United Kingdom experienced a record summer heatwave that was accompanied by plagues of 7-spot ladybirds, which took to biting humans since the drought devastated their food source of sap-sucking aphids.

LOCUST PLAGUES

African desert locusts are normally green-brown colored, solitary insects, but occasionally they undergo a dramatic change to a brightly colored yellow and black form with an alarming rate of reproduction, with simple physical contact between individuals triggering the transformation. The worst locust plagues in decades occurred between 2019 and 2022 in the Greater Horn of Africa. This was a period of intense and prolonged drought, driven by an extended La Niña event, and as the locusts stripped fields, they compounded the already serious food shortages.

↓ Locusts have a short life cycle, from egg (0 to 1 month), to immature hopper (1 to 3 months), to fully fledged adult (3 to 6 months).

↓ The Greater Horn of Africa (shaded area below) suffered devastating multi-year locust plagues around 2020.

→ Given the right weather conditions, locust populations can explode. Swarms of locusts consume lots of vegetation and devastate crops, leading to famine.

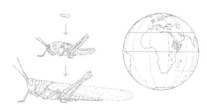

NAPOLEON AND 1812

Napoleon Bonaparte seized power in 1799 and in the first decade of the 19th century he had conducted a series of successful military campaigns across Europe, creating the French Empire and establishing a formidable reputation as a military leader.

At this time, the French and Russians were allies and had signed peace treaties, not least to keep the British at bay. However, relations were becoming strained and by 1811 the French were planning to invade Russia. Napoleon refused to comply with Russian demands to remove troops from Prussia and the campaign started in June 1812 when Napoleon's Grande Armée, consisting of half a million troops, marched quickly east into Russia.

HOT SUMMER

By July, unusually hot continental summer conditions, a war of attrition with Russian resistance, and lack of sufficient supplies slowed progress. Major losses occurred in battles over the following months and into the fall. Hundreds of thousands of troops are estimated to have been lost, but Napoleon continued toward Moscow as winter approached. At this time, St. Petersburg was the capital of Russia rather than Moscow and Russian General Mikhail Kutusov took the decision to carry out a tactical retreat from Moscow to preserve the Russian Army.

Napoleon eventually took Moscow, but it had been largely deserted and burned, so there was little in the way of supplies to fuel his army. He waited for a peace offer, occupying Moscow for a month, but it never came. Napoleon began his retreat in October as the first signs of cold weather appeared, but crucially after the Battle of Maloyaroslavets on October 24, he was forced by Kutusov to take the same route by which he had arrived in Moscow. Since this was devastated from his earlier invasion, supplies and food were scarce. Napoleon's army suffered repeated attacks, disease, and poor nutrition, and although they had suffered most of their losses due to the protracted nature of the campaign, inadequate supplies, and prolonged Russian resistance, the cold winter of 1812 was about to provide the final nail in the coffin.

↑ Intense cold and severe winter weather chipped away at Napoleon's army, which dwindled in size from hundreds of thousands to ten thousand as it marched toward Moscow (white) and retreated (shaded).

WINTER, 1812

As the French retreated, the winds turned to northerlies and November was marked by blizzards and heavy snowfall. The campaign had not planned for the extreme cold weather of the 1812 winter and both men and horses were ill-equipped for the freezing conditions. The summer uniforms of the French Army meant that mass hypothermia was now added to their list of woes. The retreat slowed and became more disorganized as the extreme cold worsened and temperatures fell below -30°C (-22°F) at the end of November. This led to further losses of tens of thousands from Napoleon's dwindling army.

A HUMILIATING RETREAT

Attacks from Russian forces to the south (also declining due to heavy losses and the extreme cold), disease, exhaustion, and the extreme cold continental weather whittled away at the army and by the time they reached western Russia in December, the army had collapsed and Napoleon was forced to return to Paris by sledge in the continuing extreme cold. This final humiliating defeat, not least due to the extreme cold of the 1812 winter, marked the beginning of the end of Napoleon's reputation.

HISTORY REPEATS ITSELF

More than a century after Napoleon, a repeat assault was launched on the Soviet Union during the Second World War. Operation Barbarossa began in the summer of 1941 on June 22, when millions of German and other Axis force soldiers attempted to take control of western Russia to secure oil supplies and forced labor.

RASPUTITSA

Early successes through the summer meant relatively rapid progress of the Axis forces and millions of Soviets were killed, with terrible attrocities committed on prisoners of war. However, resistance was already stronger than expected and in the background there were concerns about the coming weather. Hitler apparently refused to allow mention of Napoleon's failed invasion and he ignored warnings

about the "Rasputitsa," or fall mud season, when rains turn fields and tracks into deep mud. The leading German long-range weather forecaster Franz Bauer provided a forecast for a mild winter based on historical analogs. He also suggested that since the previous three winters had been colder than average and since they had never recorded four cold winters in a row, it would likely be mild.

COLD WINTER

The lack of thorough preparations for the coming weather had an enormous cost. Equipment and men were lost and progress slowed dramatically through the fall mud season. Then came the winter. A severe cold snap in the first week of December left the German troops facing Siberian winds from the north and east without adequate cold weather equipment and winter uniforms. Reported temperatures were so low (-50°C/-58°F) that uniform buttons cracked apart and German fuel solidified.

Russian troops were better prepared for the cold and the anticipated rapid collapse of Soviet lines never happened. Instead, Soviet resistance was much stronger than anticipated and Axis forces were bogged down. With the help of the weather, they were finally defeated by the Soviets in the Battle of Moscow, within sight of the Kremlin and 30 km (20 miles) of the capital. Soviet forces then retaliated and forced an extensive German retreat. The loss of troops and equipment was so severe that many historians believe this was the beginning of the end for Nazi Germany.

← The mud-season first hindered the German Army's progress toward Moscow but was then followed by extreme cold, for which they were not prepared.

1877, EL NIÑO

Only a few parts of the world had good weather observations in the 19th century. However, the limited historical observations from land and sea (recorded by trade ships), combined with written accounts from those who witnessed the weather at the time, as well as other records from tree rings, which fluctuate with temperature and rainfall, all help us to piece together a picture of this early weather.

COMPARED TO OTHER EVENTS

These early records suggest some dramatic and anomalous weather in the late 1870s. In 1877, an intense El Niño developed in the equatorial Pacific Ocean, where it warmed the tropics by at least as much as, if not more than, any of the modern big El Niño events in 1982–83, 1997–98, and 2015–16. This El Niño was also incredibly persistent and lasted from the summer of 1877 right through to the summer of 1878, extending farther west and closer to Asia than normal El Niños.

GLOBAL DROUGHT

The 1877 El Niño was at the core of an intense multiyear drought from 1875 to 1878 in Asia, Brazil, Africa, and Australia, which has been termed the Great Drought. Many of the regions affected are now well known to be at risk during El Niño and serious droughts occurred across a multitude of countries, including India, China, Egypt, Morocco, Ethiopia, southern Africa, Brazil, Colombia, and

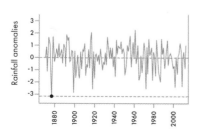

← The 1877 El Niño was at the core of a prolonged series of devastating droughts across tropical regions that included the lowest Indian summer monsoon rainfall on record.

Venezuela. India had an extensive network of rain gauges that provided one of the more complete records of the Great Drought. These show that India experienced record low rainfall in the summer monsoon season of 1877.

FAMINE

These global-reaching droughts caused mass famine between 1876 and 1878, with total human fatalities likely exceeding 50 million. The East Asian monsoon region was the worst affected and drought lasted all three years, with between 20 and 30 million fatalities. India suffered over 10 million fatalities and there were another 2 million in Brazil. It is estimated that this disaster caused around a 3 percent reduction of the global population and it has been called the "worst environmental disaster to ever befall humanity." The loss of life is comparable to that in the two World Wars and the impacts lasted for decades, exacerbating great inequalities between the tropical and extratropical regions of the world.

Understanding the Great Drought and the 1877 El Niño is as important as ever. What if it happened again? How should we respond? We know that such events are possible, but we now have over a degree of global warming on top of any future event and the impacts could be all the more severe.

GLOBAL ANOMALIES

The 1877 event was embedded in several years of unusual weather and climate. It followed a persistent multiyear La Niña event that likely built up ocean heat ready for its release during the massive El Niño that followed. It was also accompanied by an intense seesaw of ocean temperatures in the Indian Ocean, with excess warmth in the west and cooling in the east. It was then followed by unusual warmth across the whole Indian Ocean and record warmth in the North Atlantic in 1878. These latter additional factors extended and intensified the impacts.

THE 1930S DUST BOWL

By the 1920s, an influx of settlers started to farm the semi-arid American Midwest. Assuming that the recent plentiful rainfall would continue, new mechanized farming methods involved deep ploughing and the removal of soil-anchoring native plants.

DROUGHT AND DUST

In the 1930s, the climate transitioned into a decadal drought. Coupled with naïve farming methods, this reduced the soil to dust and strong westerly winds whipped up repeated intense dust storms, visibility was often near zero, and dust drifted several feet deep. Coupled with the Great Depression of the 1930s, tens of thousands of farmers were forced to give up their land in a mass exodus. We now know the Dust Bowl was at least partly the result of cooling in the tropical Pacific and warming in the Atlantic. It is a stark warning of decadal weather variations and was vividly described in John Steinbeck's famous book, *The Grapes of Wrath*.

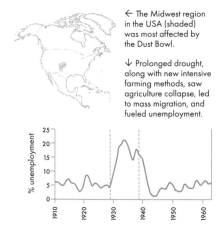

← The Midwest region in the USA (shaded) was most affected by the Dust Bowl.

↓ Prolonged drought, along with new intensive farming methods, saw agriculture collapse, led to mass migration, and fueled unemployment.

→ Destabilized surface soils led to massive dust storms like this one pictured in 1930s American Midwest. These dust storms engulfed farms and buried livestock and machinery, making it impossible for farming to continue.

ANCIENT EARTH

Weather and life are deeply connected. Life depends on the proportions of different gases in the atmosphere, but these varied dramatically in the distant past. The Earth is around 4.5 billion years old, but it was only after the first half the planet's history that primitive bacteria evolved to photosynthesize and started producing oxygen, leading to the Great Oxygenation Event.

RISING OXYGEN

The highly reactive oxygen was mostly absorbed by oceans and land at first but these eventually became saturated and oxygen in the atmosphere rose to levels comparable to today. This last rise in oxygen was followed by the Cambrian Explosion (about 500 million years ago) when life on Earth really took off, developing a multitude of oxygen-breathing forms. There were even periods when oxygen levels were significantly higher than today, which are thought to be responsible for gignatism in insects such as dragonflies with 60-cm (2-ft) wingspans and enormous millipedes around 2.5 m (8 ft) long, which have been found fossilized (see Chapter 11, pages 130–131).

↓ Oxygen levels in the atmosphere changed over the history of the Earth and it was only when they were high enough that life began.

→ Atmospheric oxygen increased dramatically just over 500 million years ago and this coincided with the Cambrian Explosion, when a multitude of strange-looking creatures evolved in the oceans.

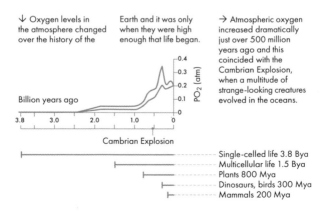

Billion years ago

PO_2 (atm)

Cambrian Explosion

Single-celled life 3.8 Bya
Multicellular life 1.5 Bya
Plants 800 Mya
Dinosaurs, birds 300 Mya
Mammals 200 Mya

THE MIGRATING MOON

While the Moon's effect on the oceans is clear from our observations of the daily tides, its effect on the weather is barely discernible from the multitude of variations caused by other factors. Apart from during a total solar eclipse, when local changes in weather occur as the Moon blots out the Sun, other lunar effects on weather are small and the monthly cycle of the Moon's waxing and waning causes only tiny changes. However, the Moon's orbit around the Earth has not always been the same and the very long-term links between the Moon and the weather are much more dramatic.

EARTH'S DISTANT PAST

Changes in the Moon's orbit are subtle and have no discernible influence on today's weather, but what if we wind time back to much earlier in the Earth's history? The Earth is around 4.5 billion years old and the Moon is thought to be about 4 billion years old. If we now think about the Moon's migrating orbit over these timescales, we can see that the effects are very large indeed. Winding right back to the point when the Moon was close to the Earth implies that the length of the day was just six hours—four times shorter than a modern day! At this time (albeit billions of years ago) there would have been enormous tides, many times the height of those today. The much more rapid rotation of the Earth at these times also had a big effect on the weather and the jet streams and rainbands we see today in the middle latitudes of the Earth would have been far closer to the Equator.

SLOW CHANGES

It turns out that the Moon is not quite in the stable orbit it appears to be. The tides themselves cause a slight torque on the Earth that results in the Moon slowly extracting spin from the Earth. This means that the Earth's spin is slowly decreasing, lengthening the day, albeit by a miniscule 2.4 thousandths of a second each century. In addition, the Moon-Earth distance is ever-so-slowly increasing, by about 3.8 cm (1 1/2 in) each year.

MILANKOVITCH CYCLES

Milutin Milankovitch (1879–1958) was a Serbian scientist who suggested that long-term changes in the orientation and distance of the Earth from the Sun as it completes its annual orbit could be responsible for changes in the weather on timescales of tens to hundreds of thousands of years and that this could trigger the onset or end of glacial periods in the Earth's weather history.

THREE CYCLES

There are three well-recognized orbital cycles that cause variations in the distribution and amount of light arriving on the Earth from the Sun. The eccentricity of the orbit describes how circular or "egg-shaped" the orbit is and this varies over 100,000 years; it's been as high as 6 percent in the past but is currently around 2 percent. Obliquity describes the tilt of the Earth's rotation axis relative to the plane of its orbit and this is currently 23.4 degrees but varies by a few degrees every 40,000 years. The precession of the Earth's axis corresponds to a wobble of the axis in a circular motion around its average position and varies over 25,000 years.

↓ Eccentricity: This is how "egg-shaped" our orbit is around the Sun. It varies on 100,000-year timescales.

↓ Obliquity: The angle of tilt of the Earth's rotation axis to the plane of its orbit around the Sun. It is now 23.4 degrees and varies on a 40,000-year timescale.

↓ Precession: This is the slow migration of the Earth's rotation axis. It rotates counterclockwise in a circle every 25,000 years.

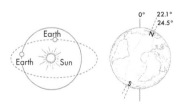

CYCLES AND BEATS

The three Milankovitch cycles come in and out of phase with each other over very long timescales to give a complicated series of beats, with peaks and troughs over hundreds of thousands of years. To make things even more complicated, there are still outstanding questions about why the Earth's climate history goes through periods with pretty regular cycling corresponding to the timescale of the 100,000-year eccentricity cycle or the 40,000-year obliquity cycle, and it's not known what causes switching between the two or why a particular cycle dominates in one period or another.

It seems likely that interaction with other parts of the Earth system, like the ice sheets, might give rise to very long timescale chaos in response to Milankovitch cycles. Having said this, although it's not as simple as a one-to-one relationship with clear cycles—for example, every 40,000 years—there is lots of evidence from ice cores and layered lake, ocean, and rock sediments that these events are an important driver of long-term changes in the weather.

OTHER PLANETS

If Milankovitch cycles affect the Earth's weather, then it seems reasonable to suggest that they might also affect weather and climate on other planets—after all, there is nothing particularly special about the Earth in this respect. Not much is known about Milankovitch cycles of the other planets, but it turns out that our near neighbor Mars undergoes much larger Milankovitch cycles than the Earth, with much greater changes in eccentricity, for example.

RECENT WARMING

One last note on Milankovitch cycles: while they're important for the long-term history of weather on Earth, they can't explain the rapid rate of warming we have seen over the last century and, in fact, the current state of the cycles would predict current cooling of the Earth. Instead, greenhouse gases are needed to account for current global warming.

DANSGAARD–OESCHGER AND BOND EVENTS

About one in every five hundred oxygen molecules is slightly heavier than the others. While most oxygen atoms contain 8 protons and 8 neutrons, the heavy oxygens contain 8 protons and 10 neutrons. This doesn't change the chemical properties and reactions of the oxygen, but it does change its physical properties enough to alter the rate of evaporation and condensation for water that contains ^{18}O from that of water with the usual ^{16}O atoms, and the difference depends on temperature.

This subtle dependence on temperature gives us a probe into ancient weather, long before the invention of thermometers. By taking cylindrical cores of Greenland ice and measuring the ratio of ^{18}O and ^{16}O it has been shown that the regional climate had large swings to warm, then cold, around every 1,500 years and that these Dansgaard–Oeschger events occurred throughout the last glacial period from around 20,000 to 100,000 years ago.

BOND EVENTS

A series of warming and cooling events is also thought to have occurred over the last 10,000 years in the more recent Holocene Epoch that followed the last glacial period. These "Bond events" have a similar period of around 1,500 years as Dansgaard–Oeschger events and they have been inferred from deposits on the Atlantic Ocean seabed, which contain cycles in the amount of material rafted by Arctic ice as it floats southward and melts.

REGIONAL CLIMATE

An interesting feature of these ancient weather events is their rapid warming over Greenland and the North Atlantic by several degrees in as little as a decade, followed by slower cooling. The effects are largest in the Northern Hemisphere and temperature changes of several degrees have been inferred from the ice core isotope records. While the effects are mainly in the Northern Hemisphere, piecing together similar records from river and lake deposits and stalactites and stalagmites in caves, as dripping water deposits minerals in banded layers, shows that these millennial events affect weather as far away as Europe and Northern and Central America.

MECHANISMS

It has been suggested the Dansgaard–Oeschger and Bond events are connected to changes in the Atlantic ocean currents that bring warm water poleward from the tropics and maintain moderate temperatures in the midlatitudes around the Atlantic basin. Others have suggested they could be related to a subtle cycling of the Sun's output over thousand-year timescales. However, there is as yet no consensus on the exact mechanism for these events.

← Greenland ice contains weather information going back hundreds of thousands of years. Subtle changes in the ratios of different isotopes of common atmospheric elements such as oxygen are affected by temperature and can be used to infer ancient conditions.

ICE AGES

I ce ages are very long periods of the Earth's history when ice was more prevalent on land and on the oceans. In fact, we are currently in an ice age—the Quaternary Ice Age, which started a few million years ago. A handful of major ice ages are known from the Earth's history and it has been suggested that at times the most severe events may have created a "Snowball Earth," where the planet was completely covered in ice right down to the Equator. The latest ice age started around 2.5 million years ago and continues today.

GLACIALS AND INTERGLACIALS

Ice ages are punctuated by glacials and interglacials where ice sheets and glaciers expand and retreat. As well as being in the Quaternary Ice Age we are also in an interglacial epoch, the Holocene. Such relatively warm interglacials like the current Holocene come and go over timescales of tens of thousands of years and the most recent glacial period retreated around 10,000 years ago.

↓ Glacials and interglacials are marked by changes in global temperature and huge variations in sea level as water is locked away in ice sheets or melts to fill the oceans.

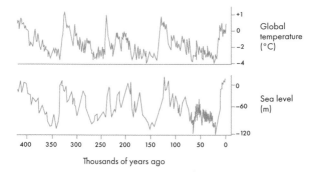

Global temperature (°C)

Sea level (m)

Thousands of years ago

PAST VARIATIONS

Ice ages and glacial periods are accompanied by expansion of the ice sheets over the polar regions and expansion of land glaciers. We know this from ice-transported deposits of boulders and rubble, which can be found well beyond the current edges of glaciers, suggesting earlier, more extensive ice fields. These also tally with ice core records and fossil pollen and other fossil evidence that show changes in flora and fauna as the ice ages come and go. All this water trapped in the ice has to come from somewhere and sea levels undergo a massive drop that can exceed 100 m (300 ft) during glacial periods. Such changes radically alter the world map and can provide land corridors for animals to migrate and colonize new areas before the interglacial warming restores sea levels.

THE MODERN ERA

Milankovitch cycles (see page 86) are thought to drive glacial-interglacial transitions within ice ages on timescales of tens to hundreds of thousands of years and the temperature changes during glacial-interglacial periods are as much as several degrees. The current phasing of Milankovitc cycles predicts that if all else were equal, we would continue in the current interglacial before heading toward a new ice age tens of thousands of years from now. However, the Earth has already warmed by more than one degree even over the last century and this rapid warming due to climate change is driving a rapid shrinking of sea ice and glaciers on land that is expected to disrupt and delay the transition to the next glacial period.

THE LITTLE ICE AGE

The Little Ice Age is not really an ice age at all but rather a cool climatic period between the 15th and 19th centuries. It is well known from historical records—for example, in Europe, where in London the River Thames would regularly freeze over, allowing ice fairs that are depicted in 19th-century paintings. The Little Ice Age contains a whole raft of extremely cold winters and these were responsible for crop failures and famines. The cause of the Little Ice Age is debated but a combination of reduction in solar irradiance and increased volcanic activity, which both cool the Earth, may have played a role.

THE LAST MILLENNIUM

Weather in the last millennium is of great interest and allows us to put recent global warming in context. A few thermometer records go back several hundred years, but data from tree rings, marine and lake sediments, and stalagmites and stalactites are needed for a global picture. Careful synthesis suggests that the "Medieval Optimum" was followed by a slow general cooling into the Little Ice Age, but it's still unclear if these were really global phenomena.

VOLCANIC FLUCTUATIONS

The clearest cause of global temperature fluctuations before the industrial period is volcanic activity. This releases sulfur dioxide, which forms sulfuric acid droplets that can last for months or years in the stratosphere, reflecting sunlight and cooling the globe. Knock-on effects on the ocean and periods of frequent eruptions, as occurred around the late 18th century, can add up to sustained cooling.

The general cooling over the last millennium ends with the sharp upturn in global temperature in the last century and recent decades are likely to have been the warmest in a thousand years.

↓ A mixture of actual temperature measurements and proxy data from other sources like tree rings or lake sediments, allow us to reconstruct millennial temperature records.

→ The Little Ice Age lasted several hundred years until the mid-19th century and was marked by frequent, severely cold winters, which are documented in books and artworks from the time.

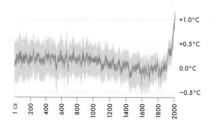

ORIGINS OF FORECASTS

Aristotle (384–322 BCE) is widely accepted as the first meteorologist. Aristotle's meteorology described the movement of vapor up into the atmosphere during hot, dry weather and back to the surface as rain during cold weather. In this sense, he was the first to discuss a hydrological cycle.

EARLY OBSERVATIONS

Explanations and predictions of the weather abound through history. A whole raft of popular proverbs and rhymes originated from shepherds, farmers, fishers, and sailors in attempts to give indicators of storms or fine weather and farming failures or bountiful crops. Many of these were based on the effects of the weather. For example, the idea that a halo around the Moon indicates wet weather on the way simply reflects that high-level clouds precede the arrival of a weather front, while the Indian proverb "The full moon grows fat on clouds" just refers to the fact that the Moon is visible when clouds are absent. Agronomical relationships such as "in the years when plums

---

IRRATIONALITY

Departures from normal weather often produced less rational explanations and consequences. For example, Egyptians believed that there were two unlucky days each month from which subsequent weather could be predicted. Whereas in 16th-century Europe there are well-documented cases of people being accused of witchcraft and burned at the stake for invoking storms and weather disasters. Ironically, the weather on Saints' days was also often thought to provide important forecasts. The weather on St. Swithen, St. Valentine, St. Eulalie, and St. Vincent's day were all said to indicate what would follow: "If the sun shine brightly on St. Vincent's day, we shall have more wine than water."

flourish, all else fails," may simply reflect weather conditions needed for different crops. Others reflect the climatology of the time—for example, "as the days grow longer, the storms grow stronger"—and describe the regular change of the weather through the year.

SCIENTIFIC RECORDS

It was not until the 17th century that consistent weather records began. Barometers started to be used in earnest as Evangelista Torricelli and Blaise Pascal realized that the air has weight. Careful measurements of the air pressure at different altitudes confirmed the idea that it was the weight of the air above a point that gives rise to air pressure. By this stage, regular thermometer measurements of atmospheric temperature were also beginning to be recorded on a regular basis. Some of these local records continue to this day, including the Central England Temperature—the longest continuous observational temperature record in the world, which started in 1659.

THE FIRST FORECASTS

During the 18th and 19th centuries, great strides were made in instrumentation and numerous weather records and actual weather forecasts appeared. The first public forecasts were issued by Admiral FitzRoy, captain of HMS *Beagle*, the ship that carried Charles Darwin around the world when collecting the evidence that gave rise to the theory of evolution. FitzRoy became convinced of the need for weather forecasts—after shipping disasters caused by bad weather, he set up a coastal observing network and issued the first shipping forecasts in 1859 and the first public weather forecasts in *The Times* newspaper in 1861.

→ The first public forecast was issued by Admiral Robert FitzRoy in *The Times* newspaper. He captained the ship that carried Charles Darwin around the world and founded the UK Meteorological Office.

L. F. RICHARDSON

Lewis Fry Richardson (1881–1953) was a physicist, mathematician, and meteorologist with a broad range of interests, including natural history as well as math and science. Richardson was a Quaker and a staunch pacifist, driving ambulances rather than entering the armed forces during the First World War, and he later carried out research on theories of war and conflict.

VISION

Richardson worked at the United Kingdom's Meteorological Office for a time and is often credited with the invention of the numerical weather forecast, although, like many scientific breakthroughs, he built on the work of others, including Norwegian meteorologist Vilhelm Bjerknes (1862–1951). Richardson imagined a vast gallery of human computers, each carrying out calculations to solve the fluid dynamical equations that govern the atmosphere for a small region of the globe, while passing results to their neighbors to solve for the weather across the whole Earth. This turned out to be a prescient vision of the numerical methods that came many years afterward with the advent of early electronic computers.

NUMERICAL FORECASTS

Richardson even made an attempt at actually implementing the method using weather observations for a particular day and performing all the calculations by hand. However, while the basis of Richardson's method was sound, in practice the implementation led to large, unrealistic pressure changes of well over 100 hectopascals (hPa) in just 24 hours and the forecast failed. Despite the poor results from this first attempt, it was Richardson's visionary idea and early attempt at forecasting that set the wheels in motion for the production of the first numerical forecasts that are the basis of today's weather forecasts. The many lives that have since been saved by modern weather forecasts and warnings of impending storms and flooding are a fitting consequence of Richardson's early ideas.

FRACTALS

In his studies of conflict, Richardson proposed that the likelihood of conflict between two countries was related to the length of their shared border. While assembling numerical data to test and establish his theories, he discovered that measuring the length of a border posed a strange problem. One might assume that as the outline of a country is measured in smaller and smaller steps, the total length of the border would converge closer and closer to the true answer. However, it didn't. Instead, it just kept increasing without limit as smaller and smaller wiggles in the border were added. This finding is an early example of a fractal and it helped inspire the subsequent development of this branch of mathematics.

↓ Richardson's vision of a vast gallery of human computers, each carrying out calculations to solve the mathematical equations of weather forecasting for a small region, is analogous to modern forecasting using supercomputers.

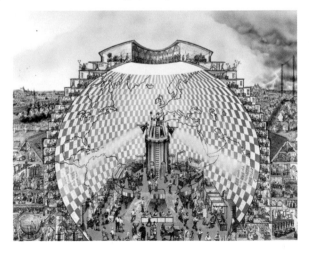

COMPUTER-BASED FORECASTS

In the 1940s, the Hungarian mathematician John Von Neumann (1903–57) instigated a weather forecasting project, applying electronic computers to the forecast problem based on his experience of developing nuclear weapons in the Manhattan Project. A team of eminent physicists, mathematicians, and meteorologists, including Jule Charney, were recruited to be involved in this milestone project. These early forecasts were produced on a state-of-the-art ENIAC computer, arguably the world's first programmable computer, using a simplified latitude-longitude version of the atmosphere with no details of how the weather was changing with height.

The early efforts of Von Neumann and his team were carried out at the Institute for Advanced Study in Princeton and demonstrated forecasts that were successful enough to stimulate similar programs elsewhere. Their success also meant that Von Neumann's project eventually led to the creation of the Geophysical Fluid Dynamics Laboratory at Princeton, which still operates as a leading center for weather and climate research today.

CONTINOUS ADVANCES

A vast network of real-time surface observations, a multitude of balloon-borne radiosondes, and, perhaps most important of all, the satellite instruments that produced regular atmospheric observations from the 1960s onward, eventually led to an explosion of weather information that could be used to estimate the state of the atmosphere. A vast array of advances have since been made to produce better physics-based models, with better mathematical schemes for solving the underlying equations and better observational starting conditions for the forecasts. Increasingly powerful computers allow the development of higher resolution computer models that now predict the future state of the atmosphere and hence the weather at ever-finer resolution and ever-longer timescales.

EARLY NUMERICAL FORECASTS

It might seem strange to try to forecast changes in the weather only in the horizontal, without taking into account the vertical. However, it's actually very reasonable for a first attempt because the horizontal scale of weather systems is much greater than the vertical. In effect, most weather systems are thin, pancake-shaped circulations in the atmosphere that are a hundred times wider than they are deep. These methods removed the instabilities that plagued L. F. Richardson's early attempt at numerical forecasting and they were able to produce plausible and skillful computer-generated forecasts by the early 1950s.

← The first computer forecasts were run by a team headed by American meteorologist Jule Charney (1917–1981). They used the first programmable computer, an ENIAC (Electronic Numerical Integrator and Computer), pictured.

CHAOS

In the 1960s, MIT researcher Edward Lorenz (1917–2008) was running computer experiments on very simple models of the atmosphere. The equations in these models couldn't produce weather forecasts like today's computer models, but they did share an important common property. Lorenz found that a tiny error in one of the variables in his model grew disproportionately as the forecast progressed and the weather in his model soon became *completely* different. He had discovered the sensitivity to initial conditions that is characteristic of chaotic systems like the weather. It means that the flap of a butterfly's wings in Brazil really can lead to a hurricane or a drought weeks later.

CHAOS AND PREDICTABILITY

While chaos means that forecasting a specific, local weather event more than a few weeks in advance from now is doomed to failure, it doesn't mean that all long-range forecasts are impossible and predictions of the type of weather in the coming months or even years are now better than ever. Neither does chaos mean the atmosphere is "random"; in fact, it's just the opposite and no matter how complicated the subsequent weather, if you start from exactly the same conditions you get exactly the same weather afterward.

↓ Small differences at the start of a weather forecast (below left) can evolve into dramatic differences just days or weeks later on (below right) due to atmospheric chaos.

→ The Lorenz model of weather is an idealized model of the weather. Despite its simplicity, it exhibits sensitivity to initial conditions whereby tiny differences on day 1 (represented by a small shift between neighboring loops in the diagram) can lead to dramatic differences later on (represented by the two wings of attraction).

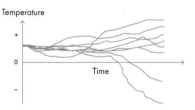

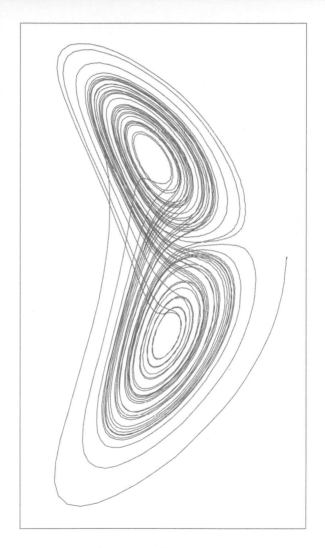

LONG-RANGE FORECASTS

Forecasting the weather well beyond the usual range of a week or so is a topic that is advancing and still the subject of much scientific research. As any well-informed meteorologist will tell you, the prospects of an accurate prediction of the weather at a particular location, on a particular day, and at a lead time of months is not possible beyond the usual progression of the seasons. The fundamental reason for this is that chaotic (unpredictable) fluctuations develop in weather forecasts due to inevitable errors in today's weather data that we use as a starting point and inevitable approximations in the computer models that we use to make the forecast. Both of these cause forecasts to diverge from the observed weather.

ENSEMBLES OF FORECASTS

To deal with this uncertainty, we make many forecasts, each containing small differences from the others so that we can assess how the whole ensemble evolves compared to the usual run of the seasons. An example is shown in the figure where ensemble forecasts in 2023 are compared to ensemble forecasts in 2020. Although individual forecasts diverge and spread out, there is still useful information here. The 2023 forecasts are evolving into a strong El Niño (see Chapter 5, page 60) and the 2020 forecasts into a La Niña. Although a precise

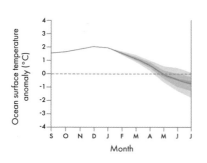

← Long-range weather forecasts become more uncertain the further ahead we go, so we use lots of forecasts to estimate the future possibilities and the range and uncertainty of future weather.

forecast cannot be made, these signals are still of great use—for example, to those making plans for water resources and potential drought or flooding in regions affected by El Niño and La Niña.

TROPICAL PREDICTABILITY

The tropics are the source of much (though not all) of this long-range weather predictability. This is because the tropical oceans are strongly coupled to the overlying atmosphere. Near the Equator, a warming of the ocean stimulates low atmospheric pressure, rising motion, and rainfall. The high predictability of the tropics means that in some regions, seasonal rainfall forecasts are as good as weather forecasts in the extratropics. In fact, it has long been known that the weather over the tropical oceans can be more persistent and easily forecast. For example, the great explorer and scientist Alexander Von Humboldt states in the 19th-century account of his journey to tropical South America that "Prognostics are also in general less uncertain on the ocean, and especially in the equinoctial parts of it . . ."

PREDICTING THE ODDS

We still haven't reached the limit for long-range weather forecasts, but they are already used in a growing number of applications where even uncertain forecasts can help predict the odds. These range from long-range hurricane predictions for insurance providers to long-range forecasts of cold snaps for energy providers that want to predict high energy demand.

EXTRATROPICAL FORECASTS

Long-range forecasts for the extratropics are much harder, but predictability of tropical rainfall has knock-on benefits for extratropical predictions. Increased rainfall over the Equator results from rising air and this air has to go somewhere. It diverges outward in the upper troposphere before it descends in neighboring regions where rainfall is suppressed. The divergence can cause the subtropical westerly jet streams to meander, triggering global-scale Rossby waves (see Chapter 11, page 140)—named after the famous meteorologist Carl Gustav Rossby—that can propagate poleward and eastward across the globe. This mechanism leads to persistent weather and long-range predictability of the weather, even in the extratropics.

ARTIFICIAL INTELLIGENCE

Artificial intelligence is perhaps a misnomer and some people prefer "machine learning" as a more measured term. Either way, these computer codes are now being used to predict the weather. This approach to weather forecasting is based on complex networks of "neurons" modeled on the structure of the brain and they show great skill in predicting the weather. These methods are not based on the fundamental physics equations that govern the fluid behavior of the atmosphere. Instead, they rely on very large datasets of past weather cases for numerical training of the algorithms. Given many thousands of past cases where the particular weather on day one is known to be followed by particular weather on day two, three, and so on, the neural network can reproduce what happens next and make successful forecasts from new data. Once set up, these methods can run and produce a new forecast almost immediately with almost no computational cost, saving energy and freeing up computing resources.

↓ Artificial neural networks contain neurons in a variety of configurations that send input signals (A) to each other, passing information along the network to produce an output (B). The signals are only transmitted if inputs exceed thresholds and this leads to complex nonlinear behavior that can be used to model and predict the weather.

→ Neurons in the real brain are much more complicated than the idealized networks we use for weather forecasting. Nonetheless, the principles are similar and the idealized neural networks used for weather forecasting are now able to make accurate weather forecasts much faster than traditional methods.

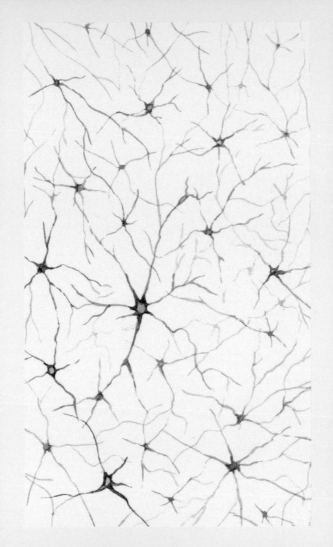

NOT A NEW IDEA

Climate change is now an overriding problem of the modern world and the issue appears daily in our news outlets. However, it is not a new idea. Climate change from greenhouse gases actually has a scientific history stretching back over 200 years.

DISCOVERY

The first widely recognized player in this story is Joseph Fourier (1768–1830), a French mathematician and physicist who is well known for his many discoveries and mathematical theorems. In the early nineteenth century, Fourier calculated the temperature of the Earth, given its distance from the Sun and the level of solar irradiance, and he showed that the Earth should be around 30°C (50°F) colder. He concluded that some other factor was at work, warming the Earth. Fourier had discovered the greenhouse effect. Scientists speculated that water vapor or carbon dioxide (CO_2) could be the culprits for the additional warming, but it was not until Eunice Foote (1819–1888) in the United States and John Tyndall (1820–1893) in the United Kingdom described thermal experiments with various candidate gases that it was established experimentally that CO_2 was likely to be the main contributor to any warming of the Earth from human activity.

We now have a much better understanding of what makes some molecules act as greenhouse gases. It is those molecules that have strong molecular vibrations in the infra-red part of the spectrum that cause the most warming and there are some molecules with hundreds or even thousands of times more potency in this respect than CO_2. Fortunately, they are very scarce in the atmosphere and are increasingly regulated to prevent emissions.

CALCULATION

The first numerical estimates of the magnitude of global warming came later in the 19th century from Svante Arrhenius (1859–1927), a Swedish chemist who carried out calculations by hand of the effects of CO_2 doubling and came up with a figure of 5–6°C (9–10.8°F). These were followed up in the early 20th century by the British engineer Guy Callendar (1898–1964), who gave revised estimates of around 2°C (3 6°F) and also provided early evidence that the world was already warming. Callendar even identified the enhanced warming in polar regions—so-called Arctic amplification—that results in the warming rate at the pole being several times greater than the globe as a whole. Our final character is Jule Charney, a brilliant meteorologist from the United States, who led a report on the likely effects of global warming. Published in 1979, it concluded that doubling CO_2 would lead to around 3°C (5.4°F) of global warming. Remarkably, all of these findings have stood the test of time and they are now validated by modern observations and computer modeling.

↓ Instruments such as this spectrophotometer revealed that gases such as CO_2, while scarce in the tmosphere, had great potential for absorbing light and warming the Earth.

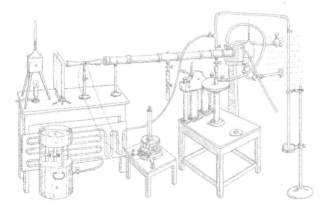

GREENHOUSE GASES

Greenhouse gases are everywhere in the atmosphere; in fact, some greenhouse gases are natural and have been present for many millions of years. Water vapor is one such natural greenhouse gas and it is present in large enough quantities to warm the Earth by around 30°C (50°F). Without this natural greenhouse gas effect, the Earth would probably be frozen solid and perhaps even uninhabitable, at least by life as we know it.

GROWING CONCENTRATIONS

We know that we are causing increases in greenhouse gases from several lines of evidence. The most direct is a set of long-running observations, where flasks of air have been regularly collected from the same site and tested for their CO_2 concentration since the late 1950s. This data is behind the famous "Keeling Curve" (after the scientist Charles David Keeling, 1928–2005), and it shows the ever-climbing level of CO_2 in the atmosphere. Other evidence includes the proportion of radioactive ^{14}C atoms in the CO_2 of the atmosphere. Radioactive ^{14}C atoms decay with time and so are much less abundant in fossil fuels. The proportions of carbon isotopes in airborne CO_2 is consistent with other estimates of fossil fuel emissions.

OTHER GASES

Methane and nitrous oxide are also increasing due to human activity. These are even stronger greenhouse gases than CO_2. Each molecule of methane (CH_4) and nitrous oxide (N_2O) produces around 80 times and 270 times more warming than each molecule of CO_2, respectively. Methane and nitrous oxide are present in smaller concentrations and are more quickly removed from the atmosphere than CO_2, but they are still important greenhouse gases.

ANTHROPOGENIC GREENHOUSE GASES

Greenhouse gases have been released into the atmosphere by human activity. Burning fossil fuels in particular, but also deforestation and other activities like cement production and transport, all release CO_2. The total amount of carbon dioxide in the air is now well over 400 parts per million, compared to around 280 parts per million in the pre-industrial period. Around 60 percent of all the carbon dioxide molecules in the atmosphere have, at some time, been released by human activities and a similar proportion of the carbon in our bodies originates from the same source!

↑ Industrial activity and deforestation has released vast quantities of greenhouse gases, primarily carbon dioxide, into the atmosphere, where it remains for decades.

GLOBAL WARMING

Global warming of the Earth's surface has been happening at an increasing rate since pre-industrial times. The Earth is currently about 1.3°C (2.3°F) warmer than in the pre-indisutrial period (defined as 1850–1900, for convenience) and most of the warming has occurred since the latter part of the 20th century. It continues at a rate of around 0.2°C (0.36°F) every decade.

RADIATIVE BALANCE

Global warming results from the trapping of infra-red (heat) radiation by greenhouse gases in the atmosphere. Infra-red is emitted by all warm bodies and it is the mechanism by which the Earth returns the energy from the Sun back into space. As greenhouse gases increase in concentration, they make the atmosphere more opaque to infra-red and the atmosphere struggles to release energy back into space. It therefore warms until it reaches a new temperature high enough to emit sufficient radiation that balance is more or less restored.

The effect of increasing greenhouse gases is equivalent to just a few Watts of power for each square meter of the Earth's surface, but this is enough to warm the globe by the amount we observe.

UNEQUIVOCAL EVIDENCE

Global warming agrees with theoretical computer models and predictions based on the laws of physics that have been made for several decades and predicted the pattern and amount of warming we are currently seeing. Similarly, although the Earth has experienced warmer periods—for example, in the Pliocene Epoch around 3 million years ago, when geological evidence suggests the Earth was several degrees warmer than it is today—the rate of recent global warming is incredibly fast. The impact of human activities in causing global warming is now considered unequivocal.

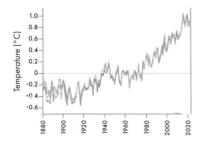

← Global temperature records produced by independent scientific research groups are in agreement: the Earth is warming well beyond the level that can be explained by natural causes and we are rapidly approaching 1.5°C (2.7°F) above pre-industrial levels.

WARMING PATTERNS

Global warming is not uniform across the Earth and it turns out that land warms faster than oceans and polar regions warm faster than the tropics. The Arctic, for example, is warming at three or four times the global average. Importantly, these patterns have been predicted for many years in experiments where CO_2 is added to the virtual Earth in our computer simulations. This adds great weight to the evidence that greenhouse gases are the cause of global warming.

The effects of global warming are collectively termed climate change. Among these, one of the big worries is sea level rise. Much of this comes from the simple fact that as we warm the oceans, the water expands and so sea levels must rise. Given that coastal regions are much more densely populated than inland regions, the potential impacts of coastal flooding, exacerbated by climate change, are enormous.

The Intergovernmental Panel on Climate Change (IPCC) was formed to determine the best scientific consensus on global warming and climate change. The IPCC has recommended thresholds to try to limit global warming to 2°C (3.6°F) and 1.5°C (2.7°F) if possible. Beyond these levels, the evidence suggests that climate change could have dangerous consequences and there are increasing chances of irreversible effects such as a shutdown of the Atlantic Meridional Overturning Circulation, irreversible damage to polar ice sheets, and perhaps even dramatic releases of methane from the ocean floor, where it is currently stored in vast quantities.

FUTURE HEATWAVES

Heatwaves invariably occur in summer and although they have a variety of definitions that vary from country to country, they always reside in the extreme upper tail of the distribution of local temperatures.

Although heatwaves are defined in terms of temperature, they cannot occur without the right pattern of winds and they almost always require a high pressure anticyclone or poleward winds to bring hot air from nearer the Equator. Anticyclones contain descending air that suppresses cloud and allows sunlight to directly heat the surface. Once the land surface dries out, sunlight can no longer be used to evaporate water and a vicious feedback of rapidly rising temperatures results.

INTENSIFICATION

Heatwaves are increasing in intensity as the planet warms and new records are set every year, with temperatures exceeding 50°C (122°F) in the United States and China in recent summers. The 2021 California wildfire and 2023 Mediterranean wildfires were both accompanied by intense heatwaves. Heatwaves also have devastating effects on people, making work impossible and leading to fatalities, especially if accompanied by high humidity.

↓ Heatwaves occur above a temperature threshold (below left). Higher temperatures increase the likelihood of heatwaves (below right) and the percentage increase in the chance of a heatwave is larger for more intense heatwaves.

→ Wildfires are a major heatwave hazard. Areas with vegetation that are prone to drought are at increasing wildfire risk.

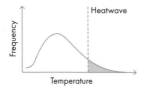

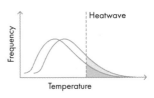

FUTURE RAINFALL

The future of Earth's rainfall is a crucial aspect of climate change that drives some of its most serious expected impacts on flooding, agriculture, and ecosystems. The picture for future rainfall builds in complexity the deeper we examine future projections of climate and rainfall trends, but a key basic fact is that warmer air contains more water vapor. This can be expressed in the well-known Clausius–Clapeyron equation, which predicts an increase of 7 percent in humidity for every degree Celsius of warming and suggests that the atmosphere will carry around 25 percent more water vapor than the current climate by the end of the century.

This increasing water vapor exerts a positive feedback on global warming and climate change because water vapor is a greenhouse gas. This amplifies the direct radiative effects of CO_2 alone. However, the extra water vapor also has to rain out of the atmosphere and this means the hydrological cycle is accelerating under climate change.

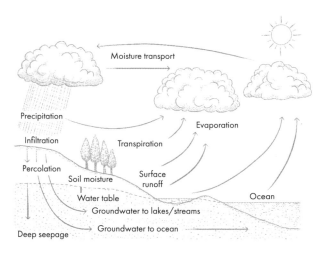

GLOBAL DIFFERENCES

Importantly, the increase in rainfall is not evenly distributed across the globe. The fact that the hydrological cycle accelerates but maintains to a reasonable degree its current pattern of wet and dry regions means that wet latitudes get wetter and dry latitudes get drier. Large rainfall increases are expected in the already wet, deep tropics near the Equator and, for example, in the Indian monsoon, as well as in the extratropical midlatitudes, where storm tracks already deliver above-average rainfall. Subtropical dry regions like the Mediterranean, southern Africa, western Australia, and parts of South America are projected to get hotter and drier, while most desert regions are set to get hotter and drier with subtropical arid regions slowly expanding poleward.

Many studies suggest that as the climate warms, storms will also dump more rainfall as they track across the Pacific and Atlantic basins and across the southern oceans. This points to another important source of impact in terms of midlatitude heavy rainfall and flooding. Again, it is the most extreme rainfall events that are expected to increase most rapidly, just as is seen for heatwaves, and this will amplify the impact of climate change as it drives unprecedented rainfall events in future.

THE PARIS AGREEMENT

Finally, we should note that the expected change in rainfall over the coming century depends on the level of future warming and this in turn depends on the level of future carbon emissions. This simple fact is driving the need to reduce CO_2 emissions to avoid the worst impacts and maintain global temperatures below the Paris recommended temperature levels of 2°C (3.6°F) and 1.5°C (2.7°F)

← The hydrological cycle is expected to accelerate with climate change since warmer air contains more water vapor and leads to increasing rainfall. The pattern of rainfall is expected to be similar in future, however, leading to already wet regions becoming wetter.

FUTURE TROPICAL STORMS

Tropical storms create some of the worst weather-related disasters and how they will change in future is a key question for the effects of climate change. Observations and future climate projections suggest with high confidence that the tropical ocean surface will warm and, all else being equal, this would increase the growth of hurricanes and typhoons. However, wind shear is also crucial for tropical cyclone development and we are less sure about how this will change in future: if it increases, then storms will be sheared out and weakened, whereas if it decreases, storms could get much stronger.

Despite the uncertainties, climate simulations and physical understanding is revealing an increasingly clear picture and we expect the strength of tropical cyclones to increase, giving more intense storms. Some scientists propose that the Saffir–Simpson scale should be extended to include unprecedented Category 6 storms. Changes in storm numbers are less clear and storms are thought to be stable or even decreasing under climate change. Either way, we expect storms to produce increasing rainfall levels and increasing storm inundation and coastal flooding as they reach land at ever-higher sea levels.

↓ There is intense discussion among meteorologists about whether we can yet see climate change in tropical cyclones. Historical records also show large decade-to-decade fluctuations.

→ Climate predictions are clear that future storms are likely to reach an unprecedented intensity, and we can expect to see more Category 4 and 5 storms in future as the Earth warms.

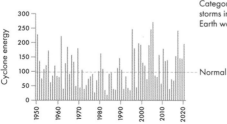

BATHTUBS AND PLUGHOLES

It is sometimes said that the direction the water in your bathtub swirls as it descends into the plughole depends on which hemisphere you are in. In the Northern Hemisphere the water is said to rotate counterclockwise, whereas in the Southern Hemisphere it is said to rotate clockwise.

This effect is due to the rotation of the Earth; because the Earth is spinning, everything on it is also inadvertently spinning, including the water in the bath. If we then put a (plug) hole in the bottom of the bath and let the water out, it is stretched into a long, thin column as it drains down the hole. This means that the water is now closer to its axis of rotation and so, in principle, it has to spin faster to conserve its total "spin," or, more formally, angular momentum, just as an ice skater spins faster if they draw in their arms. So, the bathtub idea is correct, in principle.

It is, however, false in practice. It turns out that the effect is actually very small, especially on the scale of bathtubs and demonstrations with buckets. Any residual stirring motion of the water is far more likely to govern which way the water rotates as it flows out of the hole in the bottom. So, you will see it rotate in either direction in practice. You need a very large tank, with very, very still water before it will reliably start to rotate in the sense of the Earth as the water drains from the bottom.

This spin up of vortices, also known as vortex stretching, is, however, very important on the large scale of weather systems, where it spins up the rotation of atmospheric vortices and storms with potentially devastating effects.

EFFECT OF LATITUDE

It's also worth noting that the effect of the Earth's rotation varies in strength with latitude. As with the Coriolis force, it is strongest at the poles and zero at the Equator. This means that demonstrations of the Earth's rotation for tourists at the Equator, using a rotating matchstick and a bucket of water, which is first drained on one side of the Equator and then on the other side, are sure to have been fixed by a little stirring of the water beforehand!

← The rotation of water down a plughole is like the development of rotating weather systems, but it's more sensitive to the residual movement of water than to the spin of the Earth or being in the Northern or Southern hemisphere.

CAN IT RAIN FROGS?

There are different phrases for torrential downpours. In English it "rains cats and dogs," while in Afrikaans it "rains frogs and toads." But can it really rain frogs or toads? There are no verified scientific observations, but there are numerous reports of raining small animals. Even the famous first-century Roman naturalist Pliny the Elder (23–79 CE) and the physicist André-Marie Ampère (1775–1836), of electrical fame, referred to this phenomenon.

This seems all the more physically plausible when we consider hailstones. These are often much heavier than small frogs, and yet they stay suspended aloft in the strong rising motions in deep convective storms. So, providing upward winds are strong enough, air resistance could balance the weight of small frogs.

The theory goes that small intense storms, in particular, waterspouts, whip up and send aloft groups of frogs or toads, transporting them for miles before dumping the unfortunate creatures as a "rain" of amphibians. Many reports involve aquatic animals like tadpoles and small fish, supporting the waterspout idea.

↓ Reports of raining frogs and other curious "rains" go back hundreds of years. This woodcut depicts an event from the 16th century.

↓ If the updraft in storms is strong enough it can create enough air resistance to balance the weight of small creatures like frogs.

→ Upward motion in the most intense convective rainstorms and waterspouts is strong enough to support small objects and explains the reported occurrences of raining frogs.

Weight

Air resistance

RED SKY AT NIGHT . . .

The familiar saying continues: ". . . sailors' delight. Red sky in the morning, sailors' warning." Red skies are a common indicator of the weather in many midlatitude countries, where westerly winds bring weather systems from the west. They were mentioned by Shakespeare and even described in the Bible, and it turns out there is some truth behind the saying.

The Sun's rays are made of multiple colors and the more blue the light is, the more strongly it is scattered. This means that when there is lots of dust in the air, as occurs during fine weather, blue light is scattered out of our line of sight and the sky looks intensely red.

As the Sun sets in the west in the evening and weather systems approach from the west, red sunsets therefore imply that high pressure and fine weather are on the way. Conversely, the Sun rises in the east and a red sky in the morning may imply the high pressure and fine weather have already passed and are about to be replaced by low pressure, with winds and rain.

↓ Blue light has a shorter wavelength and is more effectively scattered than red light, so if you look up at the sky, you see more blue light (B) and hence a blue sky.

→ Sunlight is composed of a spectrum of colors. Red light is less effectively scattered than blue light, so if you look toward the Sun in the evening, you see mainly red light and a red sunset.

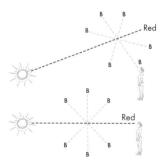

LIGHTNING NEVER STRIKES TWICE

This old adage that is regularly trotted out is definitely false and there are, in fact, many famous counterexamples where lightning regularly strikes in the same place.

Lightning occurs when the potential difference (the voltage difference) between the strike point and an electrically charged cloud overhead is big enough. To reach this point there has to be a high potential difference over a short enough distance, because it's the ratio of the two that determines the electric field strength, and a strike only occurs if the electric field strength exceeds a critical value. This is why lightning usually strikes the tallest point, because the distance to the cloud is shortest and so the electric field strength is strongest.

LIGHTNING HOTSPOTS

We can now see why it's actually very common for lightning to strike the same place many times and it is the same reason skyscrapers such as the Empire State Building are hit multiple times—sometimes even during the same storm. This is also the reason why it's dangerous to be at the top of a hill or near the only tree in otherwise flat scenery during a thunderstorm because the electric field strength between the top of the tree, or worse, the top of your head and the cloud will be a local maximum—increasing the chance of a strike.

There are verified cases of people being struck by lightning more than once. Although the majority survive the experience, it can easily result in neurological damage to the brain. The best place to be during a thunderstorm is at low altitude, well away from the highest strike points.

CONNECTION TO EARTH

Tall buildings therefore need to be earthed to allow the electricity to enter the ground where it can spread out and dissipate. This is also why electrical devices need to be disconnected during a thunderstorm to reduce any chance of the electricity from a strike running down the mains wires and causing damage to appliances or causing fire. It's also why using a wired telephone or being in the shower or bath (water is a relatively good electrical conductor) is dangerous during a thunderstorm.

← Lightning certainly does strike the same place twice and, in fact, many times in the case of skyscrapers, which provide a large electrical potential difference from nearby charged clouds.

CAN ANIMALS PREDICT
THE WEATHER?

The weather is so important to the lives of animals that it would perhaps be surprising if they had not evolved the ability to detect changes in pressure and other atmospheric conditions, so they can respond to the weather. In some cases, it is even suggested that their behavior may be used to predict a coming change in the weather.

Where there is an effect, any predictive power is often merely a response to the current weather, as in "When the glowworm lights her lamp, the air is always damp." This we know is correct from records of glowworms (glowworms are technically beetles), since the females of the species prefer to light up their display during damp weather when the night sky is cloudy, making them more easily seen by potential suitors than during a moonlit night.

Many other insects take shelter at the sign of approaching rain. Butterflies go to roost in the hedgerows and bees often return to their hive as the pressure drops and humidity rises, to avoid the rain and

QUANTITATIVE CRICKETS

All the examples given here are qualitative, in the sense that they tally with the type of weather conditions, but they don't actually measure the weather, so they can't be used to determine the actual temperature, for example. However, one famous example of an animal's response to the weather providing a quantitative measure is the cricket. Crickets chirp loudly and frequently in warm weather and the frequency of chirps is related to the temperature: the faster the chirps, the higher the temperature. According to Dolbear's Law, after American physicist Amos Dolbear (1837–1910), who monitored crickets and recorded the results, the number of chirps per 15 seconds plus 40 is meant to approximate the temperature in Fahrenheit!

high winds that are likely to follow. Similarly, the flowers of plants such as scarlet pimpernel, dandelion, and Convolvulus (bindweed) are observed to close up ahead of rain, and this is, again, likely in response to falling pressure and rising humidity so as to protect the flowers and their pollen from the damaging effects of rain.

Larger animals are also claimed to respond to the changing weather and cows laying down is said to indicate that rain is on the way. Others report that animals feed more vigorously when cold weather is on the way, but many of these reports are merely anecdotal and have not been subjected to scientific scrutiny.

POOR FORECASTERS

Many other animals exhibit apparent changes of behavior in response to the weather. Frogs are said to croak louder as rain approaches, perhaps to attract a mate during times of plentiful water for spawning. Loud singing of cicadas is said to indicate fine weather ahead (this is also a response to existing fine weather and so of limited predictive value), while leeches in bottles were used in the 19th century by George Merryweather (I kid you not) to try to predict the weather by placing them in a "Tempest Prognosticator," which was actually a glass jar with a bell that rang when one of the leeches climbed out. Fortunately, it was quickly realized that the leech prognosticator had little skill in weather forecasting . . .

→ There were serious attempts to use leeches to try to predict the weather and even a bizarre patented device, grandly titled the "Tempest Prognosticator"!

SPECIAL WEATHER DAYS?

St. Swithin was the Bishop of Winchester and lived in Anglo-Saxon times. He had asked to be buried in the open but a century later, on July 15, his body was moved. It then apparently rained for 40 days and nights as a response to the dead saint's anger, and whether it's wet or dry on St. Swithin's Day (July 15) is said to indicate the weather for the following 40 days and 40 nights.

SAINTS' DAYS

Similar to St. Swithin's Day, it is said of the European Candlemas Day—or its American equivalent, Groundhog Day (February 2)—that if the groundhog (or badger in Europe) can see its shadow, then it will be a prolonged cold winter. Equally, if it rains on February 2, then it's said that it will rain for the next 40 days. Other calendrical forecasts include St. Anders Day (November 30) in Sweden, where it is said that a wet St. Anders' Day will be followed by a cold and frozen Christmas Day. In Russia, the weather on the orthodox Christian feast day of October 1 is supposed to indicate the weather for the coming winter.

← Along with St. Swithin's Day, many days are supposed to indicate the future weather. In fact, there is no month of the year without its own special weather day.

Perhaps the most meteorologically famous examples of special weather dates are Alexander Buchan's weather "spells." Buchan was a renowned 19th-century British meteorologist and is credited with inventing the weather map, but he also noticed a series of dates when the weather seemed to suddenly depart from the usual run of the seasons. He gave key dates when spells of cold or unusual weather could be expected. However, with more data these have tended to evaporate, suggesting that Buchan's spells were simply a statistical artifact from his limited series of observations.

OTHER FOLKLORE WEATHER DAYS

The Hispanic tradition of Las Cabañuelas takes these ideas even further and uses an elaborate algorithm based on the weather on each day in January and August to predict the weather for the coming year. However, much like the other methods of forecasting using special days or other simple algorithms, there is no robust demonstration that any of these methods has real predictive skill.

ARE THEY ACCURATE?

Modern understanding tells us that all of these myths and methods are likely to fail due to the simple fact that the weather is sensitive to initial conditions: a small change today can result in big weather changes a month later and easily disrupt these simple (but appealing) rules. Although most of the folklore is simple superstition, it does illustrate a great societal need for long-range weather forecasts. There are also some small grains of truth in a few of these folkloric sayings and this stems from weather persistence. For example, the Candlemas Day/Groundhog Day folklore is not all hogwash because winter weather shows strong seasonal persistence, way beyond what would be expected from random associations of weather systems. This means that sunny, bright winter days are indeed related to persistent cold periods.

GIGANTISM IN INSECTS

During the Carboniferous Period (300–360 million years ago) there were no flowers but more primitive plants such as mosses and liverworts had evolved, and they populated vast swamps along with horsetails and ferns. These primitive plants ran riot and their decaying biomass eventually laid down the coal measures that exist today and which have been mined for their black, energy-rich deposits.

The vast forests of the Carboniferous did not just produce coal; the absorption of carbon dioxide and the production of oxygen during photosynthesis pumped up the level of oxygen in the atmosphere to 150 percent of its level today. Carboniferous life had to contend with over 30 percent atmospheric oxygen compared to the level of around 20 percent today.

↓ Gigantic insects like this huge, dragonfly-like *Meganeura* are thought to have evolved in the Carboniferous Period, a time of enhanced oxygen in the atmosphere.

AIR COMPOSITION

High oxygen levels in the Carboniferous atmosphere had all sorts of implications. The extra atmospheric oxygen made the Carboniferous weather much more dangerous; higher oxygen levels helped to fuel fires started by lightning and greatly increased the risk of forest fires. The anomalous atmosphere of the Carboniferous has even been linked to the loss of webbed feet and the development of separate digits in certain vertebrates; in other words, the development of fingers, which was a crucial evolutionary advantage for land animals. The changing atmosphere is also thought to have had a particularly dramatic effect on the size of insects and other arthropods. Small creatures like today's insects do not have lungs and instead rely on a network of small tubes (tracheoles) that permeate their body and allow passive diffusion of oxygen into their tissues. However, this process has limits and the oxygen simply can't diffuse through to great depth in the body of larger animals. This is thought to be a major reason for the current small size of insects. But the extra oxygen in the Carboniferous meant that gigantic insects could still absorb sufficient oxygen through diffusion and they were able to evolve to the gigantic proportions found in the fossil record. Other related effects have also been postulated, including the fact that excess oxygen can actually be poisonous, and so giant larvae would have been less likely to be poisoned if they were big enough to reduce their internal oxygen concentration.

GIGANTIC ARTHROPODS

Either way, the outcome of this change in atmospheric composition is clear. This high-oxygen atmosphere supported amazing creatures, including gigantic dragonfly-like insects called *Meganeura*, which hunted the skies for other insects and had a wingspan over 70 cm (2 ft). Even bigger arthropods roamed the Carboniferous forest floor, including *Arthropleura*, a gigantic millipede that was up to 2 m (6 ft) in length!

LENGTH OF DAY

The atmosphere is just a thin "skin" enveloping the surface of the Earth and its depth is no more than 1 or 2 percent of the radius of the Earth. Since the air is much less dense than the solid Earth, the total mass of the Earth's atmosphere is also much much smaller than the mass of the Earth—about a million times smaller. This all makes it hard for the atmosphere to exert a noticeable effect on the Earth.

THE ISOLATED EARTH

A second observation concerns the isolation of the atmosphere and the solid Earth as we travel around the Sun in space. Apart from small fluctuations from meteors and tidal forces from the Sun and the Moon, the Earth and its atmosphere are basically isolated. Newton's laws of physics tell us that any body isolated in this way has to conserve its "spin" or angular momentum. This means that if the Earth as a whole spins a little faster, so the day gets a little shorter, then the atmosphere must spin a little slower to compensate, and this means that the average west to east winds must change. Equally, if the average east to west winds change in the atmosphere, then the Earth's rotation and hence the length of the day, must also change to compensate.

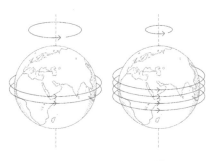

← If eastward winds strengthen, then the Earth must slow down ever so slightly to compensate and this creates a small increase in the length of the day.

MEASURING DAY LENGTH

Tiny changes in the length of day have actually been measured by astronomers. As measurements of the positions of distant stars improved, they were eventually accurate and frequent enough to detect tiny variations in the rotation rate of the Earth and hence the length of the day by variations in the speed of rotation of distant stars across the sky. We now have many decades of measurements of the Earth's rotation, and modern observations of tiny fluctuations in the length of the day, down to a few tens of millionths of a second, are possible using radio telescopes focused on quasars—very distant but powerful radio sources.

SUBTLE EFFECTS

Changes in the length of day can be detected as the atmosphere and our weather changes. Regular changes in the length of day occur with the seasons as the jet streams strengthen and weaken and irregular changes occur as the El Niño–Southern Oscillation and the Madden-Julian Oscillation flare up and dissipate. Even the high-altitude Quasi-Biennial Oscillation leaves a tiny but detectable telltale trace in the length of day. For more on these oscillations, see Chapter 5, pages 60–66.

All these variations are, of course, very small and we don't need to worry about them for our day-to-day planning of calendars and train times! They amount to fluctuations of around 1 millisecond (one-thousandth of a second) in the length of day from one year to the next, but they are important for accurate geopositioning using satellites and deep-space instruments, where small changes in the length of day can amount to tens of meters of inaccuracy as space-based instruments carry on their trajectory, oblivious to these subtle interactions between the weather and the solid Earth.

IONOSPHERE

I f we ascend above the mesosphere, we enter the upper atmosphere. Here, the extreme ultraviolet rays from the Sun are energetic enough to knock electrons off the molecules of nitrogen and oxygen to create charged ions and free electrons. The result is the ionosphere, a deep layer 80–600 km (50–370 miles) above the surface of the Earth that contains plasma—a thin soup of electrons, charged ions, and neutral molecules that conducts electricity.

The number of electrons in the ionosphere was only accurately measured with the advent of satellites, but its presence was known much earlier due to its importance for radio communications. Edward Appleton (1892–1965) was awarded a Nobel Prize for his experimental confirmation of the existence of the ionosphere in the 1920s using the fact that the layer of free electrons acts like a mirror for radio waves—reflecting signals back down to Earth and over the horizon. This allowed pioneers like Guglielmo Marconi to demonstrate the first long-distance radio transmissions.

Unfortunately, it turns out that reflecting radio waves off the ionosphere is not the most reliable communication method, since the ionized electrons vary with the 11-year solar cycle and are also easily disrupted by sudden changes in solar activity.

↓ Radio waves of the right frequency can be reflected off the electrically conducting ionosphere then transmitted to a receiver over the horizon, allowing long-distance communication.

→ Charged particles in the ionosphere collide with atmospheric atoms of nitrogen and oxygen, which then release light and give rise to the dazzling auroral displays around the Arctic and Antarctic.

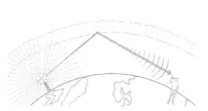

SPACE WEATHER

Space weather sounds like a strange concept. How can there be weather in space where there is no atmosphere? Space weather is the term used for variations in near-space conditions, including particles, radiation, and electromagnetic fields that mostly emanate from the Sun. These have some beautiful side effects like the aurorae, but they represent an increasing risk to society.

Indeed, while this may sound rather exotic, it can lead to some very down-to-earth and practical problems. As we become more reliant on electrical infrastructure with cables, wires, and electrical instruments, we become all the more exposed to the impacts of space weather. In March 1989, a Coronal Mass Ejection (CME) erupted energy and matter from the Sun and induced such an intense current in the power lines of the electricity distribution network in Quebec, Canada, that transformers burned out and the entire region had a blackout.

SATELLITES

Satellites, which are now crucial for communications and navigation, are particularly vulnerable, as streams of high-energy particles from solar flares and CMEs damage sensitive space-borne instruments. Heating and expansion of the upper atmosphere during a solar storm can also suddenly immerse the satellite in much denser atmosphere, leading to additional drag and disruptive orbital changes.

FORECASTS

The threat from space weather is so great that several countries monitor and forecast it. The charged particles in CMEs often take several days to arrive, providing a lead time for issuing warnings, preparing electrical infrastructure, and shielding satellites from the worst effects.

It's important we take space weather seriously, since we still haven't seen the worst the Sun can throw at us. In 1859, the so-called Carrington Event burned out telegraph wires and caused multiple fires as a massive CME impinged on the Earth. What would such an event do today with our modern, electrified infrastructure?

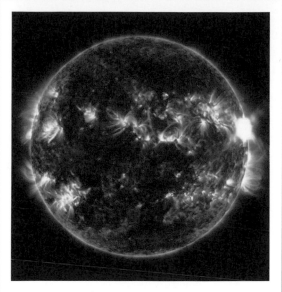

OUR STORMY STAR

Although the Sun appears to be an unchanging source of heat and light, it is, in fact, in never-ending turmoil, with fluctuating magnetic fields that carry streams of plasma—ionized atoms that have had electrons boiled off to form electrically conducting gas. Every so often, loops of magnetized plasma flare out from the Sun and sometimes whole sections break away in a Coronal Mass Ejection (CME). If this is directed at the Earth, a few days later the upper atmosphere is bombarded with charged particles and fluctuating magnetic fields that can induce damaging electrical currents at the Earth's surface.

STING JETS

S ting jets are a recently discovered feature of the most rapidly developing cyclonic storms, producing devastating wind gusts of over 160 km/h (100 mph). Cyclones contain warm and cold "conveyor belts" of rising and sinking air. Sting jets form in the descending air of the cold conveyor. After just a few hours, dense and dry air, sinking from the mid-troposphere, can accelerate to produce the pointed, hook-shaped sting jet resembling a scorpion's tale, hence the name.

Fortunately, sting jets are small, transient events just a few tens of miles across and lasting just a few hours, but they are difficult to forecast in all but the highest-resolution computer models and difficult to spot on satellite imagery. Their formation mechanism is still not fully understood, but they have been regularly documented in Atlantic storms like the great storm of 1987, which brought down an estimated 15 million trees in the United Kingdom.

↓ Midlatitude systems grow on the jet stream and evolve from ripples or waves to self-contained circulations with deep low-pressure centers.

↘ Sting jets form in the most intense midlatitude cyclones, evolving into hook-shaped cloud systems, and are still poorly understood.

→ This Atlantic storm from February 2014 contained a sting jet. Sting jets are difficult for forecasters to spot on satellite imagery.

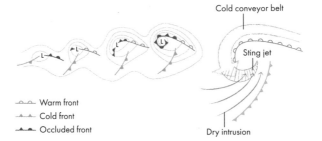

Cold conveyor belt

Sting jet

Dry intrusion

◠◠ Warm front

▲▲ Cold front

▲▲ Occluded front

GLOBAL CONNECTIONS

For a change in one part of the world's weather to cause a change elsewhere requires a transmission mechanism to carry the signal from one place to another. The fastest signals in the atmosphere are sound waves, but for weather effects to travel across the globe, a different type of wave—a Rossby wave—carries the signal through the wind. These waves are named after Carl-Gustav Rossby (1898–1957), a Swedish-American meteorologist who formally derived their properties from the mathematical equations that govern their behavior.

ROSSBY WAVE EFFECTS

Rossby showed that these waves can only travel great distances when the winds are from the west. This means that parts of the tropics, where there are easterly trade winds, are forbidden to Rossby waves. However, the other latitudes, with their westerly jet streams, provide ideal conditions for the transmission of Rossby waves. As westerly winds blow around the globe, they are deflected by mountains, producing gigantic Rossby waves and permananent weather systems in the lee of the great mountain ranges like the Rockies, the Himalayas, and the Andes.

Changes in Rossby waves can often be seen propagating across great distances, sometimes right across the globe, and can drive persistent and extreme changes in the weather. The global scale of Rossby waves means that they can drive simultaneous weather patterns in geographically remote places. In a Northern Hemisphere summer they often have around five highs and lows as they track around the globe. Simultaneous heatwaves have been caused by these circumglobal Rossby waves: they connected the devastating Russian heatwave of 2010 with the record-breaking Japanese heatwave of the same summer, and there are more recent examples in 2019, 2020, and 2022 when there were simultaneous heatwaves in the United States, Europe, and China.

ROSSBY WAVES

These atmospheric waves are on truly planetary scales. They travel across the whole globe and are refracted and reflected in a manner akin to other waves. The waves with the largest scale (the longest wavelength) travel the farthest through the atmosphere. Shorter waves are more easily refracted and they can sometimes be seen emanating from the tropics out into the midlatitudes before being reflected back into the tropics. The stationary Rossby waves that form persistent wind patterns in the atmosphere and cause prolonged heatwaves or cold snaps always propagate eastward.

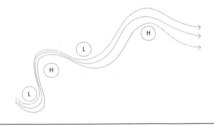

A TROPICAL CONNECTION

As well as Rossby waves triggered by mountains and circumglobal Rossby waves in summer, there are other sources of waves that originate in the slowly varying weather cycles of the tropics such as El Niño and La Niña (see Chapter 5, page 60). Tropical rainfall is strongly coupled to the tropical ocean and in most (but not all) cases, when the ocean warms, the atmosphere rises and rainfall results. The resulting disturbance in the upper troposphere can trigger Rossby waves that travel out into the rest of the globe on timescales of a week. This means that persistent tropical weather systems like El Niño result in a persistent Rossby wave. This is why El Niño and La Niña, our largest natural variations in global weather patterns, drive floods and heatwaves as far away as North America, southern Africa, and even Europe.

WHERE IS THE WEATHER HEADING?

Human beings realized they could change the global atmosphere with the development of the ozone hole (see Chapter 1, page 18). During the 1980s it was clear that policies and quick actions were needed to address this problem. Through coordinated international agreement in the Montreal Protocol, a success story was achieved. By 2000, atmospheric chlorofluorocarbons (CFCs) had leveled off and signs of ozone recovery followed.

Climate change is more difficult. It relies on the most ubiquitous gas, emitted by all human activity and, unlike CFCs, which had a few niche sources, carbon dioxide is emitted by production of one of our most basic needs: energy. On the current trajectory, carbon dioxide levels will approach 500 parts per million (ppm) by 2050 and exceed 600 ppm before the end of the century. Global climate will have warmed by several degrees and the world will be a different place. Heatwaves, coastal and pluvial flooding, and unprecedented storms will present us with a regular battery of the most challenging disasters.

The key phrase here is "on the current trajectory." Greenhouse gas emissions, the level of global warming, and the severity of future weather disasters rest in our hands

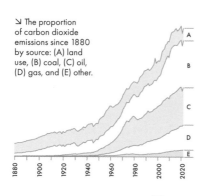

↘ The proportion of carbon dioxide emissions since 1880 by source: (A) land use, (B) coal, (C) oil, (D) gas, and (E) other.

→ Unprecedented weather is on the way as the world warms to levels not experienced by humans. More intense rainfall and storms are expected, and if greenhouse gas emissions continue at the current rate, we expect temperatures to reach levels last seen in the Pliocene several million years ago.

FUTURE FORECASTS

The idea that the laws of physics could predict the weather goes back over a century and, since that time, with the help of powerful computers to solve the intractable equations involved, the progress in forecast accuracy has been unrelenting.

COMPUTER FORECASTS

In the early decades of the 20th century, the first numerical forecasts were attempted by hand by L. F. Richardson (see Chapter 8, page 96). The development of electronic computers led to the first computer forecasts in the 1950s, and since then the accuracy of forecasts, as measured by objective numerical scores, has driven a beneficial race between forecast centers to produce the most accurate predictions. This led to remarkable progress as computers became more powerful and governments realized the important strategic and humanitarian benefits of more accurate knowledge of future weather, as was the case for the D-Day landings (see Chapter 6, page 70). Scientists pushed the limits of their forecast computer models to ever-finer resolution, included ever-more detailed physical processes, and produced ever-larger numbers of forecasts to assess the uncertainties and the range of weather possibilities.

SATELLITE OBSERVATIONS

The development of more comprehensive observations has also played a crucial role in weather forecast development. For example, the advent of global satellite observations in the 1960s and 1970s led to a massive increase in observations of the atmosphere throughout its depth and across both hemispheres, aiding the construction of better estimates of the current state of the weather from which to launch the forecast.

These developments mean that the weather forecast has improved by about 1 day every 10 years. In other words, a 5-day forecast today is as good as a 4-day forecast 10 years ago, is as good as a 3-day forecast 20 years ago, is as good as a 2-day forecast 30 years ago . . .

LONG-RANGE FORECASTING

We are now also pushing out the time range of forecasts. Although the finer local details are less predictable at longer timescales, predictions for weeks, months, seasons, and even years ahead are all now made routinely at weather forecast centers around the world. While it is the large, continental-scale patterns that are predictable on these very long timescales, this nonetheless contains useful information about the likely regional weather.

NEW TECHNIQUES

The progress has not slowed in weather forecasting since the first computer predictions over half a century ago and the future's bright for weather forecast improvements. As ever, there are new challenges and the exponential increase in the required computational cost, plus the need to reduce energy consumption, now need to be met by recoding weather forecasts to run on large numbers of parallel computer processors. An additional aid is set to come from artificial intelligence (see Chapter 8, page 104), or, more accurately, "machine learning," which can deliver thousandfold increases in the speed of forecast generation.

One last note on future weather forecasts. Human beings have a habit of underestimating what's possible by believing that their latest state-of-the-art technology has reached the limit of predictability. Hindsight shows that this often isn't the case, and although we have theoretical estimates, we still don't know the ultimate limits of weather forecasts.

→ Forecasting the weather is a triumph of science: from the first simple observations using barometers, thermometers, and anemometers, we now use supercomputers and satellites to provide ever-increasing accuracy and forecast range.

GEOENGINEERING

Geoengineering is a radical and controversial idea that some argue may be necessary to combat climate change by deliberately intervening in the weather and climate. The most direct methods call for the removal of greenhouse gases from the atmosphere, while indirect methods propose to negate greenhouse gas effects through interventions like reducing the amount of sunlight reaching the Earth's surface. While some of this sounds like science fiction, these methods have been suggested as a plausible last resort if greenhouse gas emissions are not reduced and climate change causes intolerable weather disasters.

GREENHOUSE GAS REMOVAL

Simple methods of greenhouse gas removal include afforestation by planting trees, which absorb carbon dioxide from the atmosphere, fixing it in the wood they produce. Various mechanical or chemical means of direct carbon capture from the air have also been proposed. The carbon dioxide is then stored deep underground or under the ocean bed. Other proposals include encouraging the growth of phytoplankton in the ocean by adding dissolved fertilizing compounds to the surface ocean. Phytoplankton then bloom and absorb carbon

---

ETHICAL ISSUES

There are big problems associated with geoengineering. Some of these are ethical: who should be in control of the deployment of such an approach? How can we be sure about the effects and side effects? What should we do if some regions benefit while others suffer? Given all these problems, some have called for international regulation for the public good or even an international non-use agreement, and there is a clear need for strict governance and public disclosure of research results.

---

dioxide before dying and sinking to the ocean bed, carrying the gas with them. Yet other methods rely on agriculture to create biochar—a partially combusted organic material that can be left in the ground—or bioenergy with carbon capture and sequestration, which relies on partial combustion of crops to produce energy, followed again by storage of the residual carbon dioxide.

SOLAR RADIATION MANAGEMENT

Solar radiation management is a rapidly deployable but controversial method of geoengineering. It's been proposed that a small fraction of the light arriving from the Sun be reflected back into space, cooling the planet to counter global warming. Injection of aerosol material into the stratosphere to mimic the natural reflective cooling effect of a volcanic eruption, marine cloud brightening using artificial cloud seeding, and even space-based reflectors placed between the Earth and the Sun to reflect incoming sunlight have all been suggested. While these methods are ingenious, they pose potential side effects. For example, the stratospheric aerosol method comes with a risk of increased hurricane activity, reduced Indian monsoon rainfall, and unintended warming over northern Europe and the United States.

WEATHER MODIFICATION

A third method of geoengineering that is less talked about is the possibility of weather modification through intervention prior to extreme weather. In this case, ensembles of weather forecasts can be interrogated to find the small differences that increase the risk of an impending weather disaster. In principle, interventions such as Solar Radiation Management could then be used to reduce the risk.

→ If climate change impacts become too costly, artificial cloud brightening (A), stratospheric aerosols (B), or even space-borne mirrors (C) could reduce solar radiation and rebalance temperatures, but all are fraught with side effects and ethical issues.

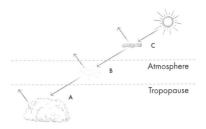

WEATHER ON OTHER PLANETS

Weather on other planets is a fascinating area for new discoveries. Although the atmospheres on all planets follow the same laws of science, they each have different sizes, sit at different distances from their star, have different rotation rates, and different atmospheric constituents, and this leads to some bizarre and unexpected weather.

MARS

Mars has a tenuous atmosphere, less than 1 percent of that on Earth, and it is mostly (95 percent) carbon dioxide. This leads to some novel effects in which the carbon dioxide freezes and falls as "snow" in winter! Although the surface terrain suggests that water was present on Mars billions of years ago, there is no liquid water today and so Mars has no rain. However, it does have cyclones and gigantic dust storms and, although the average temperature is a freezing -60°C (-75°F), it is a comfortable 20°C (70°F) at the Equator.

VENUS

Venus is far hotter than Earth, with a much thicker atmosphere, which is, again, mainly carbon dioxide. This drives an intense greenhouse effect so that Venus has the hottest surface temperature of any planet in the Solar System. The atmosphere of Venus also exhibits an unusual superrotation, where the atmosphere spins faster than the planet itself, and the planet is shrouded in clouds that rain down hot sulfuric acid.

→ We can learn a lot by studying the atmospheres of other planets. Even with their varying features, such as different rotation rates, they still show storms and jet streams and are important test cases for understanding weather.

GAS GIANTS

Jupiter and Saturn are gas giants with no solid surface and they have greater mass and deeper atmospheres than the other planets in our Solar System. They are mainly composed of hydrogen but have a troposphere and a stratosphere just like Earth. They also have jet streams and cyclones and the equivalent of a Quasi-Biennial Oscillation (see Chapter 5, page 66) but with a much longer period of around four years for Jupiter and even longer for Saturn.

EXOPLANETS

We are just starting to learn about the weather on planets outside our Solar System. It seems that many of these newly found planets, or at least the ones currently detectable by astronomers, fall into the same class of "hot Jupiters." These are often tidally locked, with the same side of the planet facing their local star at all times and reaching very high temperatures of thousands of degrees.

THE END OF WEATHER

At some point in the distant future the weather will end. There will be no more cyclones and rainfall, no more heatwaves and hailstorms, and along with any lifeforms that have persisted, the weather on Earth will end.

THE EVOLVING SUN

The Sun is about 4.5 billion years old, but its energy output is increasing by about 1 percent every hundred million years. This means that in a billion years, the energy reaching the Earth from the Sun will be so intense that the oceans will simply boil away. This will likely mark the end of any complex life remaining on Earth at that time.

If that were not enough, the Sun will also eventually come to a dramatic end itself. In a few billion years, it will burn out its hydrogen fuel, no longer producing any energy from the nuclear fusion of hydrogen to helium. It then has to succumb to the irresistible force of gravity, collapsing before exploding into a red giant star. This event will expand solar material outward, enveloping the inner planets and bombarding the Earth with particles and radiation that will sweep away any residual atmosphere once and for all.

↓ The Sun is a fairly common type of star. Through astronomical observations and theoretical calculations we can infer its evolution. It has billions of years of life left (it is currently a yellow dwarf) but will eventually balloon into a red giant.

→ By the time the Sun dies, Earth's climate will be unbearable for life and the oceans will have vaporized. The ocean and atmosphere will be swept into space and weather will end.

Planetary nebula	Low-mass star	Yellow dwarf	Red giant	Stellar nebula	White dwarf

GLOSSARY

advection
Transport that results from being carried by the wind.

aerosol
Small suspended particle or droplet in the atmosphere.

angular momentum
The amount of spin of a body in rotational motion. It is conserved in the absence of tangential force. Formally, it is the product of rotational velocity, mass, and the square of the distance from the rotation axis.

anticyclone
A high pressure weather system that circulates clockwise/counterclockwise in the Northern/Southern hemispheres.

Atlantic Meridional Overturning Circulation
The net northward flow of the upper Atlantic Ocean and net southward flow of the deep Atlantic Ocean that transports heat and salinity poleward.

centrifugal force
The outward force on a body that is traveling in a curved motion.

climatology
The average weather conditions over many years—30 years are commonly used.

convection
The rising motion of pockets of warm air, often responsible for cloud and rain.

Coriolis force
An apparent force due to the rotation of the Earth that is proportional to speed and acts to the right/left of the direction of travel in the Northern/Southern hemispheres.

cyclone
A low pressure weather system that circulates counterclockwise/clockwise in the Northern/Southern hemispheres.

ENSO (El Niño–Southern Oscillation)
The natural interannual variability of the tropical Pacific climate, which warms/cools the equatorial ocean in its El Niño/La Niña phase and has knock-on effects worldwide.

front
The boundary between two different air masses. This can be sharp, with an almost discontinuous jump in temperature. A cold front is a boundary between cold and warm air where the cold air is advancing, a warm front is a boundary between cold and warm air where the warm air is advancing, and an occluded front is where the warm air is cut off from the surface and there is no longer a warm front at ground level.

Fujita scale
Tornado reference scale from F1 (greater than 117 km/h/73 mph, weakest) to F5 (greater than 420 km/h/ 261 mph, strongest).

Hadley Cell
The rising of air in the deep tropics, poleward flow in the upper troposphere, descent on the edge of the tropics, and return flow toward the Equator in the lower troposphere.

lapse rate
The rate at which temperature changes with height.

MJO (Madden–Julian Oscillation)
Prominent weather fluctuations that travel across the West Pacific every 40 to 60 days.

mesosphere
The third and upper layer of the neutral atmosphere at around 50–85 km (30–50 miles) altitude.

monsoon
Regular seasonal change in wind direction accompanied by heavy rainfall.

NAO (North Atlantic Oscillation)
The seesaw of pressure between Iceland and the Azores that is the largest single factor determining seasonal weather around the North Atlantic.

ozone layer
The region of high ozone concentration in the lower half of the stratosphere.

polar vortex
An intense cyclonic circulation system in the winter polar stratosphere, which contains very cold stratospheric air.

pressure
The force per unit area exerted by the air. In the atmosphere, the pressure decreases exponentially with altitude, decreasing by a factor of three roughly every 7 km (4.5 miles).

QBO (Quasi–Biennial Oscillation)
The natural variation of winds in the equatorial stratosphere that oscillates from easterly to westerly and back to easterly on average every 28 months.

radiosonde
Balloon-borne weather recording instrument that ascends through the atmosphere, recording and transmitting atmospheric measurements as it ascends.

refraction
The bending of the path taken by a wave as it passes through a different background medium.

Rossby wave
A planetary-scale wave that can propagate great distances across the globe.

Saffir–Simpson scale
Hurricane reference scale from 1 (119–153 km/h/74–95 mph, weakest) to 5 (greater than 253 km/h/ 157 mph, strongest).

storm surge
The temporary rise in sea level associated with a cyclonic storm. This is separate from the astronomical tides but is most dangerous when the storm surge coincides with a spring high tide.

stratosphere
The second layer of the atmosphere at around 10–50 km (6–30 miles) altitude.

tropics
The region between 30°N and 30°S of the Equator.

tropopause
The layer between the troposphere and stratosphere. To some extent it acts as a lid on tropospheric weather systems.

troposphere
The lower layer of the atmosphere, which contains most of our weather from the surface to around 10 km (6 miles) altitude.

FURTHER READING

Reports and Books
Assessment Reports of the Intergovernmental Panel on Climate Change: https://www.ipcc.ch/assessment-report/

Bohren, C. F. 2003. *Clouds in a Glass of Beer: Simple Experiments in Atmospheric Physics.* Dover Publications Inc.

Emanuel, K. 2005. *Divine Wind: The History and Science of Hurricanes.* Oxford University Press.

Gleick, J. 1999. *Chaos: The Amazing Science of the Unpredictable.* Vintage.

Lamb, H. H. 1995. *Climate, History and the Modern World.* Routledge.

Norbury, J., and I. Roulstone. 2013. *Invisible in the Storm: The Role of Mathematics in Understanding Weather.* Princeton University Press.

Philander, S. G. 2006. *Our Affair with El Nino: How We Transformed an Enchanting Peruvian Current into a Global Climate Hazard.* Princeton University Press.

The Royal Meteorological Society: Weather A–Z. 2022. Natural History Museum.

Royal Society Review of Geoengineering: https://royalsociety.org/topics-policy/publications/2009/geoengineering-climate/

Scaife, A. (ed.). 2016. *30-Second Meteorology: The 50 most significant events and phenomena, each explained in half a minute.* Ivy Press.

Weather: A Force of Nature. 2022. Royal Meteorological Society.

Academic Books
Hoskins, B. J., and I. N. James. 2014. *Fluid Dynamics of the Midlatitude Atmosphere.* Royal Meteorological Society, Wiley.

Houghton, J. 2002. *The Physics of Atmospheres,* 3rd Edition. Cambridge University Press.

James, I. N. 1998. *Introduction to Circulating Atmospheres.* Cambridge University Press.

Vallis, G. K. 2017. *Atmospheric and Oceanic Fluid Dynamics.* Cambridge University Press.

Articles and Online Resources

Baldwin, M. et al. 2001. "The Quasi–Biennial Oscillation." Reviews of Geophysics, American Geophysical Union.

Bauer, P., A. Thorpe, and G. Brunet. 2015. "The Quiet Revolution of Numerical Weather Prediction." Nature, 525: 47–55.

The Biblical Locust Plagues of 2020. BBC: https://www.bbc.com/future/article/20200806-the-biblical-east-african-locust-plagues-of-2020.

Brönnimann, S. 2005. "The Global Climate Anomaly 1941–1942." Weather, 60: 336–342. https://doi.org/10.1256/wea.248.04.

European Centre for Medium Range Weather Forecasts: https://www.ecmwf.int/.

Forecasts of the El Niño–Southern Oscillation. From the International Research Institute for Climate and Society: https://iri.columbia.edu/our-expertise/climate/forecasts/enso/current/.

Hurrell, J. et al. 2003. "An Overview of the North Atlantic Oscillation." Geophysical Monograph Series, American Geophysical Union.

Martínez-Alvarado, O. et al. 2012. "Sting jets in intense winter North-Atlantic windstorms." Environmental Research Letters.

The Met Office: https://www.metoffice.gov.uk/ contains Weather educational pages: https://www.metoffice.gov.uk/weather/learn-about/weather.

"Mount Tambora and the Year without a Summer." University Corporation for Atmospheric Research: https://scied.ucar.edu/learning-zone/how-climate-works/mount-tambora-and-year-without-summer.

National Meteorological Library: https://artsandculture.google.com/project/met-office.

National Oceanic and Atmospheric Administration: Scientific agency in the USA with much weather, climate, and oceanography information, www.noaa.gov.

National Weather Service, USA: https://www.weather.gov/ contains many Weather educational pages: https://www.weather.gov/education/resources.

NOAA National Severe Storms Laboratory: Tornadoes FAQ, https://www.nssl.noaa.gov/education/svrwx101/tornadoes/faq/.

Scaife, A. et al. "What is the El Niño–Southern Oscillation?", Weather, 74: 250–251. https://doi.org/10.1002/wea.3404.

Space Weather Prediction Center at NOAA: https://www.swpc.noaa.gov/.

United Nations Environment Programme, Ozone Secretariat, https://ozone.unep.org/.

Zhang, C. 2005. "Madden–Julian Oscillation." Review of Geophysics, American Geophysical Union.

INDEX

ACKNOWLEDGMENTS

The author is indebted to the Met Office and the University of Exeter for the opportunities they provided to work on weather and climate science. Thanks also to my wife Samantha and son Isaac, who patiently reviewed draft sections of *The Little Book of Weather* as they were being written.

Acknowledgments for illustration references: 24 Data from the NASA Goddard Institute for Space Studies; 27 After data from the India Meteorological Department; 33 After *Encyclopedia of Atmospheric Sciences*. Eds North, Pyle, and Zhang, 2015; 63 After Skinner et al, *Weather*, 2022; 75 Martin Grandjean; 78 After Singh et al., *J. Climate*, 2018; 82 Heinrich D. Holland; 92 RCraig09; 107 Royal Society of Chemistry Environmental Chemistry Group; 114 After Zhang et al. ACIAR, 2002; 116 After U.S. Env. Prot. Agency; 120 Universal History Archive/UIG/Shutterstock; 138 After Schultz and Vaughan, *Bull. Amer. Met. Soc.*, 2011; 142 Efbrazil.

ABOUT THE AUTHOR

Adam Scaife leads research and production of long-range weather forecasts at the UK's Met Office and is a professor at the University of Exeter. He is the author of *30-Second Meteorology* and was recently awarded the Edward Appleton Medal by the Institute of Physics and the Royal Meteorological Society's Buchan Prize for his pioneering work on understanding, simulating, and predicting the weather and climate.

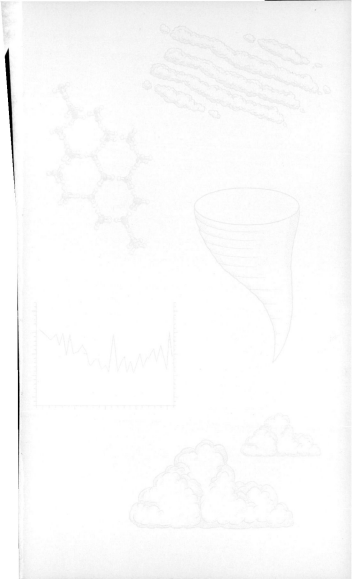